83

Sewers and Sewage Works

By the same author

Sewerage Engineering
Regional Planning
Surface Drainage
Sewerage Design and Specification
The Municipal Engineer
Sewage Treatment: Design and Specification
A Code for Sewerage Practice
Surface-water Sewerage
Building Sanitation
Rifle and Gun
Rifleman and Pistolman
Sewerage and Sewage Disposal
The Work of the Public Health Engineer (with Dr S. F. Rich)
The Small Cellar
Pumping Station Equipment and Design
Design of Surface-water Sewers
Public Health Engineering Practice
Escritts' Tables (with V. P. Escritt)

Sewers and Sewage Works

with Metric Calculations and Formulae

by
L. B. Escritt

London
GEORGE ALLEN & UNWIN LTD
Ruskin House   Museum Street

First published in 1971

This book is copyright under the Berne Convention. All rights reserved. Apart from any fair dealing for the purpose of private study, research, criticism or review, as permitted under the Copyright Act, 1956, no part of this publication may be reproduced, stored in a retrieval system, or transmitted, in any form or by any means, electronic, electrical, chemical, mechanical, optical, photocopying, recording or otherwise, without the prior permission of the copyright owner. Enquiries should be addressed to the Publishers.

© L. B. Escritt, 1971

ISBN 0 04 628002 2

PRINTED IN GREAT BRITAIN
in 10 point Times New Roman
BY ALDEN & MOWBRAY LTD, OXFORD

To Sidney and Erna Rich

Preface

The change from Imperial to metric units has set a problem for those who write on engineering subjects for, while a date for complete change-over has been stated, there is good reason to believe that imperial weights and measures will continue in use for many years.

Another problem is the extent to which the various recommendations of the several metrication committees should be accepted. In some quarters there is a tendency to agree with anything suggested by any official or semi-official body. It is only too easy to shelve one's responsibilities and slavishly follow recommendations, even if this means bowing to the absurd; but proper selection of units suitable to the quantities that they have to measure is important, for it has much to do with the saving of time and money.

A unit of measurement should be smaller than the quantity it has to measure. It would not, for example, be reasonable to measure the height of the average man in furlongs. Nevertheless, a recommendation has been made which is still less reasonable: it has been suggested that flows of water and sewage should be measured in cubic metres per second.

Just how impracticable this is becomes clear when one considers what is the order of magnitude of soil-sewage flows. If one excludes the gigantic sewage works that serve the Greater London Council, Glasgow, Birmingham and Manchester and the very large works serving Sheffield, Nottingham, Bradford and west Herts, there does not appear to be one sewerage authority in Great Britain that can claim that it has a sewage-treatment plant receiving a dry-weather flow of as much as a single cubic metre per second. How then can this unit be used for measuring the much smaller flows received by the multitude of lesser sewage works and in all the branch sewers for which nearly all flow calculations have to be made?

A more reasonable suggestion was that flows should be measured in litres per second. But when one is considering flows in and out of tanks and reservoirs, it is common sense that the capacities of the tanks should be measured in the same units as the flows in the pipes: if the flow is in litres per second, the capacity of the reservoir should be in litres. However, the Littleton reservoir of the Metropolitan Water Board contains nearly as many litres as there are centimetres of distance between the earth and the moon. One would never measure such distances in centimetres: why then give reservoir capacities in litres?

Careful consideration brings to light that the only metric equivalents that can give convenient figures for measuring domestic water and sewage flows and the capacities of reservoirs and sewage tanks are cubic metres per *minute* and cubic metres. These are the units used in this book.

Some committees have recommended that the centimetre, square centimetre and cubic centimetre, decimetre and square decimetre should never be used, and they have also laid down that the second should be the only unit of time, the minute being abolished. It has even been suggested that the hour and day should be done away with. This setting of restrictions on mathematics would appear to be purposeless: it is certainly harmful and, for this reason, should be opposed. Engineers should use every mathematical tool at their disposal: engineering is a serious occupation, not a conventional game.

There is good reason to retain the minute, for the second is unsuitable for measuring long periods of time such as times of concentration of river or sewerage catchments, sewage-tank detention periods and the long detention periods of sludge digestion tanks. These times are measured in minutes, hours, days, weeks and months and should continue to be so measured. In fact, a public health engineer need never consider the second at all in any of his calculations.

For the above reasons the author has selected what he considers to be the most practical units in preparing the calculations and formulae herein, in the belief that these will be used by the majority of readers to their advantage. Those who wish to conform to any committee pronouncements can easily make conversions to the units they prefer.

Sewers and Sewage Works is a new textbook giving, in metric values, all the main calculations and formulae required by those engaged on the design of sewers and sewage-treatment works.

L.B.E.

Contents

PREFACE	page 7
PART I: SEWERS	
1. General Principles of Sewerage	17
2. Calculating Sewer Sizes	25
3. Sewage Pumping	48
4. Pneumatic Lifting of Sewage and Sludge	69
5. Construction of Sewers	75
6. Sewer Appurtenances	89
7. Coastal and Other Special Problems	100
PART II: SEWAGE WORKS	
8. The Nature of Sewage and its Measurement	121
9. General Requirements of Sewage Treatment	133
10. Preliminary Treatment of Sewage	142
11. Separation and Treatment of Storm Water	154
12. Sedimentation	163
13. Sludge Disposal	176
14. Sludge Digestion	187
15. Percolating-filter Treatment	202
16. Activated-sludge Treatment	214
17. Production of Effluents of Extra High Standard	228
18. Sewage Treatment for Isolated Buildings	232
APPENDIX: IMPERIAL AND METRIC EQUIVALENT UNITS	248
BIBLIOGRAPHY	250
INDEX	252

Tables

1. Approximate Minimum Gradients for Drains and Sewers — page 20
2. Vertical Distances when Clay Pipe Sewers Cross — 22
3. Vertical Distances when Concrete Pipe Sewers Cross — 23
4. Soil-sewer Calculation Sheet — 28
5. Impermeability Factors — 30
6. Intensities of Rainstorms — 33
7. Typical Lloyd-Davies Calculation — 34
8. Run-off Coefficients — 37
9. Typical 'Rational' Method Calculation — 38
10. Run-off per Roofed or Paved Hectare — 39
11. Simplified 'Rational' Method Calculation — 40
12. Data for Time–Area Graph — 43
13. Economic Diameters of Pipework and Valves — 55
14. Starting Times of Electric Pumps — 61
15. Proportional Depths, Areas and Velocities of Circular Culverts Flowing partly Full — 61
16. Approximate Loss of Head through Fittings — 63
17. Loss of Head through Reflux Valves — 64
18. Efficiencies and Power Factors of Electric Motors — 65
19. Calculation of Optimum Diameter of Rising Main — 65
20. Repayment of Loans — 66
21. Compression of Air — 70
22. Efficiency of Air Lifts — 71
23. Performance of Air Lifts — 72
24. Efficiencies of Low-pressure Air Blowers — 73
25. Vitrified Clay Pipes with Flexible Joints — 76
26. Maximum Depths of Brick-barrel Sewers — 82
27. Dimensions of Cast-iron Segmental Sewers — 84
28. Air and Water Tests for Sewers and Drains — 87
29. Thicknesses of Manhole Walls — 94
30. Dangerous Gases and Vapours Liable to be Present in Sewers — 97

12 TABLES

31. Booking for Tidal Experiments	page	104
32. Discharge Coefficients for Sea Outfalls		107
33. Square Roots of Numbers		108
34. Areas of Land Required for Land Treatment		136
35. Minimum Gradients for Sludge Mains		140
36. Settlement Speeds of Siliceous Particles		143
37. Working Conditions for Disintegrators		147
38. Effect of Screen Spaces on Quantity of Screenings		148
39. Spill-over from Storm Tanks		160
40. Effect of Flocculation		164
41. Specific Gravity and Specific Heat of Sludge		178
42. Comparative Costs of Sludge-disposal Methods		179
43. Reciprocals of Surface Coefficients		193
44. Thermal Conductivities of Materials		194
45. Discharge of Gas through Orifices		200
46. Loads on Percolating Filters		206
47. Recommended Sizes of Air Mains for Diffused-air Works		220
48. Delivery of Chain Pumps		241
49. Proportions of Septic Tanks		241
50. Minimum Capacities of Small Percolating Filters		243

Formulae

1. Scobey: Velocities in Pipes	page	18
2. Scobey: Discharges of Pipes		18
3. Scobey: Discharges of Pipes		18
4. Lloyd-Davies: Storm-water Run-off		27
5. Impermeability Factors for Housing Estates		31
6. Bilham: Rainfall Intensities		31
7. Ministry of Health: Rainfall Intensities		32
8. Ministry of Health: Rainfall Intensities		32
9. 'Rational' Formula: Storm-water Run-off		36
10. Change of Impermeability during Rainfall		36
11. Simplified 'Rational' Formula: Storm-water Run-off		41
12. Basis of Tangent Method		43
13. Specific Speeds of Centrifugal Pumps		50
14. Sizes of Centrifugal Pumps		53
15. Dimensions of Centrifugal Pumps		54
16. Dimensions of Centrifugal Pumps		54
17. Dimensions of Centrifugal Pumps		54
18. Dimensions of Centrifugal Pumps		54
19. Dimensions of Valves		55
20. Dimensions of Valves		55
21. Dimensions of Valves		55
22. Diameters of Valve or Penstock Handwheels		57
23. Time Required to Open or Close Sluice Valves		58
24. Pull on Valve or Penstock Spindles		58
25. Storm-water Storage		62
26. Velocity Head		63
27. Power Demands for Pumping		64
28. Maintenance Costs of Pumping Stations		65
29. Air Requirements of Sewage Ejectors		71
30. Maximum Spans for Cast-iron Pipes		80
31. Earth Pressures on Sewers		81
32. Widths of Cascades		93

14   FORMULAE

33. Discharges of Tanks via Outfall Pipes   page 106
34. Discharges of Tanks via Outfall Pipes   106
35. Discharges of Tanks (approximation)   107
36. Effective Head on Sea Outfall   112
37. Height of Waves   113
38. McGowan: Strength of Sewage   123
39. Discharge of V-notch Weirs   130
40. Discharge of Rectangular Weirs   130
41. Discharge of Standing-wave Flumes   131
42. Land Required for Sewage-treatment Works   134
43. Cost of Sewage-treatment Works   135
44. Mogden Formula: Charges for Trade Wastes   139
45. Flows through Vertical Screens   146
46. Discharge of Standing-wave Flumes   151
47. Widths of Constant-velocity Detritus Channels   151
48. Widths of Constant-velocity Detritus Channels   152
49. Settlement of Suspended Solids   165
50. Capacities of 60° Pyramids   171
51. Crimp: Change of Volume of Sludge   177
52. Loss of Heat through Walls, General Formula   193
53. Loss of Heat through Gasholders   193
54. Loss of Heat through Vertical Walls   194
55. Danson & Jenkins: Oxygen Uptake of Activated Sludge 215
56. Performance of Activated-sludge Works   216
57. Detention Period for Activated-sludge Treatment   218
58. Mohlman: Sludge-volume Index   226
59. Sludge Age   226

PART I

SEWERS

1

General Principles of Sewerage

Flow in sewers

Sewers, other than rising mains and inverted siphons, are designed as if they were open watercourses in that surcharge is not permitted, the maximum capacity of the sewer being considered its discharge when it is just filled to soffit level. Thus the hydraulic gradient of the sewer running full is the fall of the soffit in the direction of flow. At changes of diameter and at junctions, sewers are, therefore, usually arranged so that the crown of the incoming sewer is not lower than that of the outgoing.

At low rates of flow, when the sewers are not filled, the hydraulic gradient approximates to the gradient of the invert. Thus some advantage as to velocity at low rates of flow may be gained by arranging for incoming invert to be at outgoing invert level at changes of diameter, bringing incoming crown below crown of outgoing sewer. In such cases the hydraulic gradient of the sewer running full is what it would be were incoming and outgoing crowns level, and not the actual gradient of the invert or crown.

Many formulae have been given for flow in pipes, culverts and channels, some of historical interest, some theoretically satisfying because they can be applied to fluids of all densities and viscosities and some purely empiric being based on careful examination of numbers of tests on flow of water or sewage in pipes, culverts or channels. In the first two classes are included some formulae which involve very large errors (e.g. in the region of 20% to 40%) which rule them out for practical use to engineers. The empiric formulae tend to be much more accurate, but each formula is strictly accurate when applied to pipes, culverts, or open channels of the particular material and particular standard of workmanship on which it was based.

In practice it is often necessary to vary the materials of construction. Thus, formulae used in the design of sewers are mostly general formulae applied regardless of the material or construction. An example is Crimp and Bruges's formula which, over wide ranges of sewer diameter and gradient, involves moderate error, being somewhat pessimistic about the discharge of small pipes, optimistic as regards flows in large-diameter culverts and rather too optimistic about flows in large open channels or culverts running partly full.

Of the pipe-flow formulae in recent use, Scobey's formula [23] for flow in concrete pipes laid to average standards of workmanship is probably as good as any for the design of vitrified clay or concrete pipe sewers. (The modified formula used in *Escritts' Tables*, which is in close agreement with Scobey's formula over the range of diameters to which the latter applies, is intended to give greater accuracy when used for determining flows in very small pipes or very large culverts.)

Scobey's formula reads

$$V = \frac{26 \cdot 459 D^{5/8}}{I^{\frac{1}{2}}}, \tag{1}$$

$$Q = \frac{0 \cdot 00002078 D^{21/8}}{I^{\frac{1}{2}}}, \tag{2}$$

$$L = \frac{0 \cdot 0003463 D^{21/8}}{I^{\frac{1}{2}}}, \tag{3}$$

where D = diameter in millimetres,
V = velocity in metres per minute,
Q = discharge in cubic metres per minute (values can be calculated from Tables 32 and 33),
L = discharge in litres per second,
I = length divided by fall.*

There is some recent evidence that the formulae applicable to pipes or culverts running full are not accurate when applied to open channels or pipes or culverts of the same material but running partly full. It appears that, probably because of imperfections of gradient, open channels or culverts running partly full discharge less than the theoretical amount, the reduction in discharge varying according to the irregularity of the actual as compared with the theoretical

* Length is length of pipe measured down the gradient and is not identical with the horizontal distance although it usually approximates to this in sewerage work.

gradient. The author has suggested that, on the basis of such data as are at present available, flows in channels should be assumed to be 17% less than as calculated by the pipe-flow formula applicable to the same material.

Gradients

Sewers are designed not only to accommodate the estimated flows, but to ensure velocities of flow sufficient to prevent silting. Soil sewers generally should be laid to gradients sufficient to promote a velocity of not less than about 46 metres per minute when the sewer is full. Combined sewers also are usually self-cleansing when laid to gradients steep enough to produce this velocity.

Surface-water sewers are often designed so as to flow at about 46 metres per minute when full, but this velocity is not always adequate to remove heavy silt.

It is not recommended that any *maximum* velocity should be laid down as applicable to all circumstances, because the detrimental effects of high velocities are very varied according to the concentration of grit in the sewage, the diameter of the sewer and the materials of construction. Attempts to limit velocity by restriction of gradient have, in the past, involved excessive cost at doubtful advantage, and there has been little agreement as to what the maximum velocity or gradient should be.

The recommendation is that vitrified clay, concrete and cast-iron pipe sewers, particularly those of small size and seldom flowing full, should be laid at gradients not less steep than dictated by the falls of the land. Inverts of large sewers should be formed of vitrified brickwork or equally hard material, to protect against scour.

The solids in sewage include the lighter, organic materials, some of large size, which should be held in suspension by the turbulence of flow, and the heavier mineral detritus. The detritus consists of particles of gravel etc., which slide down the invert of the sewer causing scour, and sands and silts which may or may not be in suspension, according to circumstances. Just what velocity of flow will keep the sewer clean and unobstructed depends on the quantities of heavy mineral matter and light organic matter, and the quality of construction of the sewer. Thus the velocity that is satisfactory in one instance may not be so in another.

However, most engineers both in Great Britain and in America

TABLE 1. *Approximate minimum gradients for drains and sewers*

Diameter (millimetres)	Gradient (Length/fall)	Diameter (millimetres)	Gradient (Length/fall)
100	105	1000	1906
125	138	1050	1995
150	174	1075	2084
175	209	1125	2188
200	251	1150	2291
225	288	1200	2399
250	331	1250	2512
275	380	1350	2754
300	417	1425	2884
325	457	1500	3162
350	501	1575	3311
375	550	1650	3467
400	603	1725	3802
425	661	1800	3981
450	692	1875	4169
475	759	1950	4365
500	794	2025	4571
525	832	2100	4786
550	912	2175	5012
575	955	2250	5248
600	1000	2325	5496
625	1047	2400	5754
650	1097	2550	6026
675	1148	2625	6310
700	1202	2700	6607
725	1259	2850	6918
750	1318	2925	7244
775	1380	3000	7586
800	1446	3150	7943
825	1514	3225	8318
875	1585	3375	8710
900	1660	3450	9120
925	1738	3675	9550
975	1820	3750	10000

agree that if a sewer or drain* is designed so that the velocity in it is at about 46 metres per minute when that sewer or drain is running full and there is usually a goodly flow at some time during the day, silting or stoppage should not occur. Table 1 gives the approximate minimum gradients for drains and sewers or hydraulic gradients for rising mains that will ensure self-cleansing conditions.

General requirements

One hundred millimetres diameter is the minimum for any sewer or sewage rising main. Public sewers are rarely less than 150 millimetres diameter. All sewers of 150 to 750 millimetres internal diameter should be laid in straight lines between manholes and there should be no changes of diameter or gradient except at manholes. This applies to public sewers and to private sewers, whether or not they are laid in the street.

Manholes should be provided at all changes of diameter, direction and gradient. Manholes on private and public sewers should not be more than 90 metres or 110 metres apart respectively. Manholes should be placed at all junctions on public sewers, but there need not be a manhole at the junction between a lateral private sewer or drain and any other sewer or drain provided a manhole is placed on the lateral within 12 metres of the public or private sewer or drain to which it is connected. The junction between a drain and a sewer should be such that the flow enters obliquely to the direction of flow in the sewer to which connexion is made.

When sewers exceed 750 millimetres in diameter they need not be laid in straight lines between manholes. Nevertheless the limiting distances between manholes should not be exceeded. The practice of some engineers of widely spacing manholes on the larger sewers is inadvisable, as it adds to the dangers to which men working in sewers are exposed.

To protect them from damage, sewers should be laid with at least 1·2 metres of cover from crown to ground level, when in roads, and with at least 0·9 metre of cover elsewhere. They should be sufficiently deep to pick up all drain connexions, but not unnecessarily deep involving excessive expenditure. To find the depth necessary to pick

---

* A sewer is a drain which drains more than one curtilage. There may be private or public sewers but the public ownership of a drain does not make it a public sewer.

up any proposed drain connexion from a new property, allowance should be made for a fall of not less than 1 in 105 from 0·6 metre below ground level at the point of discharge of the most distant proposed gully or soil pipe to 0·3 metre above the crown of the proposed sewer. When a sewer is laid to pick up existing premises a fall of at least 1 in 105 should be taken from the invert of the nearest inspection chamber to 0·3 metre above the crown of the sewer.

Location of sewers

Sewers in the highway should be so located that the manholes come either in the dead centre of the carriageway or in the footway or verge, not in the centre of a traffic lane. Where separate sewers are provided, soil and surface-water sewers should be arranged to facilitate drain connexions and, where the land on one side of the road is lower than that on the other, the lower sewer should be on the downhill side of the road. It is usual for the soil sewer to be lower than the surface-water sewer.

Usually, but not always, junctions between sewers at road junctions etc., make it necessary for the soil sewers to be set sufficiently below the surface-water sewers to permit cross-over. Apart from this often necessary difference of level, the greatest economy of construction and ease of drain connexion is effected by laying the sewers with equal amount of cover or, as above described, with slightly more cover for the sewer on the downhill side of the road, which should preferably be the soil sewer.

TABLE 2. *Minimum vertical distance in metres from invert to invert to ensure clearance between outsides of sockets when vitrified clay pipe sewers cross one another*

Diameter of lower sewer in millimetres	Diameter of upper sewer in millimetres				
	150	225	300	375	450
150	0·256	0·265	0·284	0·299	0·311
225	0·342	0·351	0·369	0·384	0·397
300	0·436	0·445	0·464	0·479	0·491
375	0·528	0·537	0·556	0·570	0·582
450	0·616	0·625	0·644	0·659	0·671

TABLE 3. *Minimum vertical distance in metres from invert to invert to ensure clearance between outsides of sockets when concrete pipe sewers cross one another*

Diameter of lower sewer in millimetres	Diameter of upper sewer in millimetres								
	150	225	300	375	450	525	600	675	750
150	0·290	0·299	0·306	0·320	0·329	0·339	0·344	0·351	0·357
225	0·375	0·384	0·400	0·406	0·415	0·424	0·430	0·437	0·442
300	0·467	0·476	0·488	0·493	0·503	0·513	0·518	0·527	0·538
375	0·549	0·558	0·570	0·580	0·589	0·598	0·604	0·610	0·616
450	0·634	0·645	0·655	0·665	0·675	0·683	0·690	0·696	0·701
525	0·720	0·729	0·741	0·750	0·760	0·769	0·774	0·781	0·787
600	0·803	0·812	0·824	0·833	0·842	0·850	0·857	0·864	0·870
675	0·885	0·895	0·909	0·915	0·925	0·935	0·940	0·945	0·951
750	0·966	0·976	0·991	0·996	1·005	1·015	1·020	1·026	1·036

Frequently soil and surface-water sewers are laid at different levels in the same trench or in closely adjacent trenches, in which case concrete support is necessary to prevent the upper sewer being let down by the earth under it subsiding.

The orthodox method of soil sewerage for new housing estates, of laying the sewer in the road, is the best to adopt for all estates of detached or semi-detached houses, for it gives each householder the advantage of having no drains or sewers on his land other than those for which he alone is responsible. The method, however, is comparatively costly, and it is sometimes more economical as regards capital costs to lay the soil sewer through the gardens behind houses, although this may lead to troubles later.

Sewers laid behind houses, whether in public or private ownership, differ neither technically nor legally from those laid in the road, and they must not be treated as if they were drains. They must be laid in straight lines between manholes and there should be no connexions to them other than from the last manhole or inspection chamber of the drainage system of each individual curtilage. No interceptors are permitted on these or any other sewers: no restriction can be made on the number of houses which can be connected to these or to any other sewers.

Sewerage systems

Sewers are laid according to three systems, the combined, the separate and the partially-separate systems. In the combined system, soil sewage and surface water are accommodated in the same sewers, which are designed on the basis of surface-water run-off, the soil flow, by comparison, being negligible in quantity. In the separate system, soil sewage is discharged to the soil sewers and conveyed to the sewage-treatment works, while surface-water is accommodated in entirely separate surface-water sewers which discharge to the nearest watercourses. The partially-separate system is a term applied to separate systems in which some surface-water, usually that from back yards and back roofs, is discharged to the soil sewers by permission of the local authority.

2

Calculating Sewer Sizes

Sewer sizes are calculated according to estimated peak flows, available hydraulic gradients and suitable hydraulic-flow formulae. There are two types of flow to be estimated, first the flow of soil sewage from domestic and industrial premises, and second, the run-off of rainfall.

In the design of soil sewers the soil-sewage flow, plus some infiltration, has to be considered. In the design of surface-water sewers, rainfall run-off only has to be considered. This also applies to combined sewers, for the peak quantity of soil-sewage flow is usually so small in relation to the peak storm flow that it can be neglected. In the design of partially separate sewers taking both soil and surface-water flows, the soil flow may or may not be taken into account according to whether or not it is negligible compared with the storm flow.

Capacities of soil sewers*

Soil sewage is mostly water that has been supplied by the water authority but which, having been used for domestic and other purposes, has become sullied and needs to be removed. For this reason there is some similarity between the hourly and daily rates of flow of company's water and of sewage and, therefore, knowledge of water consumption is always valuable when designing sewers and sewage works.

There are differences. Part of the company's water is lost by evaporation, soakage into the ground or drainage by means other than soil or combined sewers. When water is used for watering crops or roads, or for car washing, building works, etc., it may not reach the sewers, and there is always some leakage from water mains to the

* The American term is 'sanitary' sewers.

subsoil. On the other hand, infiltration to sewers from the subsoil usually exceeds water-supply losses, with the result that, except in limestone areas where leakage exceeds infiltration, the flow in sewers generally tends to exceed the water supply by very varying amounts. A fair average infiltration for an old sewerage system could, perhaps, be one-fifth of the dry-weather flow.

Water demands and sewage flows vary from one locality to another according to the amount of industry, the extent to which domestic premises have baths and other sanitary provision, and other factors. In modern housing estates the water demand is about 0·125 cubic metre per person per day. The demand for day schools, offices, shops and industrial premises (excluding process water) is usually taken as 0·045 cubic metre per head per day. A demand of 0·23 cubic metre is usual for mental and children's hospitals and 0·36 cubic metre for general hospitals.

It has been usual to allow a water demand of 0·045 to 0·07 cubic metre per head per day for villages with main water supply, 0·12 for small provincial towns and 0·15 for most moderate-sized towns. But water demands and sewage flows are tending to increase noticeably and, at the present time, the rate of flow to local government sewage-treatment works serving populations of 10000 persons upwards tends to average in the region of 0·25 cubic metre dry-weather flow per head of population per day. By 'dry-weather flow' is meant the average daily rate of flow as recorded during a dry period of preferably 21 days during which the amount of rainfall has never exceeded 2·54 millimetres in any one day.

The rate of flow of soil sewage varies throughout the day from being little more than the rate of infiltration early in the morning to about twice the average daily rate an hour or so before midday in the case of a small drainage area, or one and a half times the average daily rate at about 4 p.m. in a very large sewerage area. To allow for variation of rate of flow during the day and variation from day to day, it is usual to design separate soil sewers to take, when running full, not less than four times the average dry-weather flow and, not infrequently, the sewers are designed to take six times the average dry-weather flow as a precaution against unforeseen additional flows.

To find the flow in any one sewer or branch sewer, common practice is to determine the flow per house from the flow per head of population, and to count the number of houses on a large-scale map.

There may be some difficulty in determining the number of persons per house for this figure has, of recent years, tended to decrease in new housing estates but increase where old houses have been converted to flats. Perhaps the most useful guide can be the number of persons per water service pipe served by the water authority. The figures in Greater London, which are very representative, are 2·7 persons per household but 4·45 persons per water service pipe.

Thus, suppose that the dry-weather flow is 0·12 cubic metre per head per day, there are $4\frac{1}{2}$ persons per house and the separate soil sewers are being designed to accommodate, when running full, four times dry-weather flow, then the design flow will be 0·0015 cubic metre per minute per house. A typical calculation will then be as set out in Table 4.

Capacities of surface-water and combined sewers

The approach to surface-water sewer design is quite different from that of soil-sewer design for, whereas the flow in soil sewers is virtually in direct proportion to the number of houses or head of population served, surface-water sewers have to accommodate the run-off of rainfall, the intensity of which reduces as the drainage area increases.

At one time it was not uncommon for rates of rainfall to be allowed according to the area of the catchment. A typical example was the Leeds scale in which rates of rainfall varying from 1 inch (25·4 mm) to $\frac{1}{4}$ inch (6·35 mm) per hour were allowed for areas of 20 acres (8·1 hectares) to 600 acres (243 hectares).

The reason why higher intensities of rainfall had to be allowed when making drainage calculations for small areas than for larger areas was the observation that the storm of any particular frequency of occurrence that produces the greatest flow in the sewer is the one that has a duration about equal to the time of concentration, i.e. the time taken for the storm water to arrive from the most distant part of the catchment to the point at which flow is desired to be known.

Thus, out of early 'flat rate' practice developed the 'time-of-concentration' methods, the Lloyd-Davies method [13] in Great Britain and the 'rational' method in America.

In the Lloyd-Davies method, run-off from a drainage area is calculated according to a formula which, converted to metric units, reads

$$Q = 0\cdot 16 A p R, \tag{4}$$

TABLE 4. *Soil-sewer calculation sheet*

Location, manhole to manhole	Number of houses	Peak flow (cubic metres per minute)	Gradient (Length/fall)	Required diameter of sewer (millimetres)	Discharge (cubic metres per minute)
A5 to A4	80	0·12	145	150	0·89
A4 to A3	600	0·9	229	225	2·05
A3 to A2	1200	1·8	380	300	3·39
A2 to A1	1275	1·91	417	300	3·24
B3 to B2	80	0·12	159	150	0·85
B2 to B1	450	0·675	174	150	0·81
B1 to A3	520	0·78	275	225	1·87

where Q = peak rate of flow in the sewer at the end of the time of concentration, in cubic metres per minute,
Ap = impermeable area, or gross area multiplied by the impermeability factor, in hectares,
R = intensity of rainfall applicable to the time of concentration, in millimetres per hour.

The Lloyd-Davies method involves a procedure of trial, for the time of concentration is not known until the sewer has been designed. Thus, the designer has to assume a time of concentration, base a rainfall intensity on this, calculate by the formula the rate of flow, and determine by flow formula (or tables) the diameter of sewer required to take this flow at the available gradient. He then finds the velocity of flow and divides the length of the sewer by this to find the time of concentration, to which quotient he may add a few minutes 'time of entry' to allow for flow over the surface of the ground to the sewers. The new time of concentration gives a new rainfall intensity and, with this, the calculation is repeated. This procedure is continued until there is no further change in the rainfall intensity.

Impermeable area

The impermeable area can be found by multiplying parts of the gross drainage area by impermeability factors. There are impermeability factors for various types of surface (see Table 5) which were accepted prior to recent research and which still appear to be accurate means of determining 'ultimate impermeability' or ratio of total run-off to total rainfall. The impermeability of an area can be found by measuring the extent of each kind of surface and completing the calculation thus:

Sample area	square metres	square metres
Roofs of houses	90 000 × 0·90 =	81 000
Asphalted carriageways	150 000 × 0·80 =	120 000
Footways, etc.	60 000 × 0·65 =	39 000
Garden land	625 000 × 0·00 =	zero
Total	925 000	240 000
Impermeability factor	$\dfrac{240\,000}{925\,000} = 0\cdot26$	

Then impermeability factor multiplied by gross area in hectares gives impermeable area in hectares.

It will be noted that in the first half of Table 5 are surfaces of high

TABLE 5. *Impermeability factors*

Type of surface	Factor
Metal, glazed tile and slate roofs (Fruhling)	0·95
Watertight roof surfaces (Kuichling and Bryant)	0·70 to 0·95
Ordinary tile and roofing papers (Fruhling)	0·90
Asphalt and other smooth and dense pavements (Fruhling)	0·85 to 0·90
Asphalt pavements in good order (Kuichling and Bryant)	0·85 to 0·90
Closely jointed wood or stone-block pavements (Fruhling)	0·80 to 0·85
Stone, brick and wood-block pavements with tightly cemented joints (Kuichling and Bryant)	0·75 to 0·85
Block pavements with wide joints (Fruhling)	0·50 to 0·70
Stone, brick and wood-block pavements with open or uncemented joints (Kuichling and Bryant)	0·50 to 0·70
Average of all above	**0·806**
Cobble-stone pavements (Fruhling)	0·40 to 0·50
Inferior wood-block pavements with open joints (Kuichling and Bryant)	0·40 to 0·50
Macadam roadways (Kuichling and Bryant)	0·25 to 0·60
Macadam roadways (Fruhling)	0·25 to 0·45
Gravel roadways and walks (Kuichling and Bryant)	0·15 to 0·30
Parks, gardens, lawns, meadows, depending on surface, slope and character of subsoil (Kuichling and Bryant)	0·05 to 0·25
Wooded areas, depending as before (Kuichling and Bryant)	0·01 to 0·20

impermeability which include roof and road materials of good quality and which have an average impermeability of about 0·8, and below these are surfaces of low impermeability, in particular, parks, gardens, meadows and woodlands; in the above calculation, however, garden land has been taken as having zero impermeability. The reason for this is that rainfall runs off roofs and paved areas so much earlier than it runs off gardens, meadows and woods that the run-off from the latter type of surface comes too late to affect the peak run-off and can be neglected in the majority of calculations for developed areas: to include it usually leads to error. In statistically analysing the recorded run-offs from a number of developed areas in which the rainfall had also been recorded, the author found that the

CALCULATING SEWER SIZES

impermeability could be expressed most accurately and consistently by assuming that all roofed and paved areas had an impermeability factor of 0·8 and all gardens, lawns and meadows an impermeability factor of nil. Consequently he recommends the adoption of this rule in those circumstances where roofed and road surfaces can be measured easily.

There are, of course, exceptions. Where a large area is covered almost entirely by factory roofs, a factor of 0·95 could well be adopted, or where an area of grassland steeply slopes down so as to discharge rainwater onto a road, a proportion of the grassland run-off should be allowed for.

For estimating the impermeability factor for housing estates, the following formula may be used

$$\text{Impermeability factor} = 0.064 N^{\frac{1}{2}} \qquad (5)$$

where N is the number of houses per hectare.

This rule, which was based on two separate surveys made in Greater London, is not applicable to some modern housing estates with overhead pedestrian walkways and other exceptionally large paved areas.

Rainfall intensities

The formula most generally accepted for determining rainfall intensities in Great Britain is Bilham's formula [3] which, converted to a convenient metric form, reads

$$R = \frac{267 \cdot 7 N^{0 \cdot 2817}}{t^{0 \cdot 7183}} - \frac{152 \cdot 4}{t}, \qquad (6)$$

where R = millimetres of rainfall per hour,
N = number of years between storms of this magnitude,
t = duration of storm in minutes.

From this formula can be calculated the intensity of rainfall to be expected on the average during a storm of any particular duration and frequency of occurrence.

In sewer design the duration of rainfall is taken as being equal to the time of concentration of the drainage area: the engineer has to decide how *infrequent* a storm of this duration should be allowed for in the design. Surface-water sewer design is always a compromise between extravagance for the sake of a high degree of safety and parsimony with considerable risk of flooding. Reasonable practice is to spend sufficient money to make it unlikely that anything more than minor damage will be caused by floods during the life of the sewer. It would not be reasonable to spend vast sums of money so as to ensure that no flooding at all would occur during a period of several hundred years.

Over the years there have been some changes of practice as to the frequency of storm that should be assumed. But experience has shown that, if the storm likely to occur once every 3 years is assumed together with proper allowances for impermeability of surfaces and other factors that determine the difference between rainfall and run-off to sewers, results will almost invariably be satisfactory without extravagance having been involved. On occasions of phenomenal rainfall combined with some other unusual circumstances, there will be floods. But these will be extremely rare: the author has known of no more than three minor floods occurring after the construction of new drains or sewers to recognized standards in 46 years of surface-water sewer design. If such floods did not occur sometimes, the fact would be evidence of general over-design of drains and sewers.

The reasons that less frequent storms do not have to be considered are

1. Surcharge of the sewers, particularly those of small size in the upper part of the catchment, permits considerable increases of capacity without causing flooding.
2. Storage in manholes on small sewers holds up flow and reduces peak discharges.
3. Storms of high intensity do not cover the whole of large catchments at their maximum intensity except on extremely rare occasions.

Rainfall intensities of storms liable to occur on the average once in 3 years are given in Table 6.

Table 7 gives a typical Lloyd-Davies calculation as it would have been made in the early 1930s, but the figures have been altered to conform to metric units. The rainfall intensities have been calculated according to the old Ministry of Health formulae [17] which converted to metric units, read

$$R = \frac{762}{t+10} \quad \text{for storms of up to 20 minutes' duration,} \tag{7}$$

$$R = \frac{1016}{t+20} \quad \text{for storms of 20 to 100 minutes' duration,} \tag{8}$$

where R = rainfall intensity in millimetres per hour,
t = duration in minutes.

TABLE 6. *Intensities of rainstorms liable to occur once in 3 years*

Duration (t) (minutes)	Intensity (R) (millimetres per hour)	Duration (minutes)	Intensity (millimetres per hour)	Duration (minutes)	Intensity (millimetres per hour)
11	51·3	32	25·5	110	11·1
12	48·5	34	24·5	120	10·4
13	46·1	36	23·6	130	9·9
14	43·9	38	22·7	140	9·4
15	42·0	40	22·0	150	9·0
16	40·3	42	21·2	160	8·6
17	38·7	44	20·6	170	8·2
18	37·3	46	20·0	180	7·9
19	36·0	48	19·4	190	7·6
20	34·8	50	18·9	200	7·3
21	33·7	55	17·7	210	7·1
22	32·7	60	16·7	220	6·9
23	31·8	65	15·8	230	6·7
24	30·8	70	15·1	240	6·5
25	30·1	75	14·4	250	6·3
26	29·3	80	13·8	260	6·1
27	28·6	85	13·2	270	6·0
28	27·8	90	12·7	280	5·8
29	27·3	95	12·2	290	5·7
30	26·6	100	11·8	300	5·6

The rational method

The Lloyd-Davies method did not come into general use for some 30 years after it had been formulated, by which time some modifications had been introduced for the intended purpose of increasing its accuracy but which actually led to sewers being oversized in the majority of instances.

These modifications were designed because it had been observed that, with the rainfall formulae then in use, it was theoretically possible for the run-off from part of a drainage area, under the rain-

TABLE 7. *Typical Lloyd-Davies calculation*

Gross area (hectares)	Impermeability factor	Impermeable area (hectares)	Time of concentration (minutes)	Rainfall (millimetres per hour)	Run-off (cubic metres per minute)	Length divided by fall	Required diameter of sewer (nominal millimetres)	Capacity (cubic metres per minute)	Velocity (metres per minute)	Length (metres)	Time of flow (minutes)
2·0	0·26	0·52	5+5 min time of entry = 10	38·1	3·30	350	300	3·7	50	244	5
4·5	0·26	1·17	10+7 = 17	28·2	5·5	360	375	6·5	57	387	7
15·0	0·26	3·90	17+3 = 20	25·4	16·5	325	525	16·6	74	220	3
20·0	0·26	5·20	20+3 = 23	23·6	20·4	500	675	25·9	70	215	3

fall intensity applicable to the time-of-concentration of that part, to be greater than the run-off from the whole area under the rainfall intensity applicable to the total time of concentration. For example, should a complete catchment have an impermeable area of 4 hectares and time of concentration 20 minutes, the run-off according to a Lloyd-Davies calculation and the Ministry of Health rainfall Formulae 7 and 8 would be

$$0.1\dot{6} \times 4 \times \frac{1016}{20+20} = 16.9 \text{ m}^3/\text{min}.$$

But should part only of the catchment have an impermeable area of 3·5 hectares and time of concentration of 11 minutes the run-off from the part area would be

$$0.1\dot{6} \times 3.5 \times \frac{762}{11+10} = 21.2 \text{ m}^3/\text{min}.$$

With the assumptions that statistical rainfall intensity varies with duration but impermeability remains constant, the above is sound. But the fact is that while rainfall reduces in intensity with its duration, impermeability increases, and to such an extent that the effect of irregular distribution of impermeable area is greatly reduced. If, in the above example, we use in lieu of the Ministry of Health Formula the statistical rainfall figures given in Table 6 and apply to the impermeable area the run-off figures in Table 8 (which divided by 0·8 give values as calculated by Formula 10), the calculation for the whole area becomes

$$0.1\dot{6} \times 4 \times 34.8 \times \frac{0.535}{0.8} = 15.5 \text{ m}^3/\text{min},$$

and that for the part area

$$0.1\dot{6} \times 3.5 \times 51.3 \times \frac{0.406}{0.8} = 15.2 \text{ m}^3/\text{min}.$$

Because proper allowance was not made for change of impermeability during rainfall, sewers were designed to discharge on the average about 50% more than necessary during the period between the two World Wars. It was when this became obvious to experienced engineers that researches were made bringing to light the importance of change of impermeability.

Not only were the modifications of the Lloyd-Davies method generally liable to lead to oversizing of sewers, but some were mathematically unsound, even ridiculous, and most were unduly complicated. The complication had reached such a point that, after the Second World War, calculations which, if properly executed, could often have been completed in a matter of minutes, were being programmed into electronic computers. Moreover, in some instances the errors concealed by this method of calculation led to sewers being sized three to four times larger than was necessary.

As compared with this practice the American 'rational' methods do not suffer from theoretical complication and, being empirically based, tend towards accuracy. For this reason, after having studied as much British data as are available, the writer recommends that one of the best of the American 'rational' methods (for there are several versions) should be used in conjunction with an appropriate British run-off coefficient. In this variety of the 'rational' method [10], run-off is calculated according to a formula (very similar to the Lloyd-Davies formula) which, when converted to metric units, reads

$$Q = 0.16 \, CRA, \qquad (9)$$

where Q = run-off of storm water to the sewers in cubic metres per minute,
C = run-off coefficient,
R = rainfall intensity in millimetres per hour,
A = roofed and paved part of drainage area in hectares.

The values of C are determined empirically for any particular locality and, like the figures for rainfall, vary with the time of concentration. They include for all factors such as impermeability, change of impermeability, average irregularity of distribution of impermeable area, time of entry, etc. To arrive at a figure applicable to Great Britain the author found from the previously mentioned data that change of impermeability could be expressed by the formula

Average impermeability during rainfall

$$= i\left(\frac{t-2+4.6052 \log_{10}(2/t)}{t}\right) \qquad (10)$$

where i = ultimate impermeability factor,
t = time of concentration in minutes.

With this formula and taking i as being 0·8 for all paved and roofed areas, the figures in Table 8 were derived.

CALCULATING SEWER SIZES 37

TABLE 8. *Run-off coefficients for roofs and pavements of developed areas in Great Britain*

Duration of storm (minutes)	Run-off coefficient c	Duration of storm (minutes)	Run-off coefficient	Duration of storm (minutes)	Run-off coefficient
11	0·406	32	0·611	110	0·727
12	0·427	34	0·620	120	0·732
13	0·446	36	0·627	130	0·736
14	0·463	38	0·634	140	0·740
15	0·478	40	0·640	150	0·743
16	0·492	42	0·646	160	0·746
17	0·504	44	0·651	170	0·749
18	0·515	46	0·656	180	0·751
19	0·526	48	0·661	190	0·753
20	0·535	50	0·665	200	0·755
21	0·545	55	0·674	210	0·757
22	0·553	60	0·682	220	0·759
23	0·560	65	0·690	230	0·760
24	0·567	70	0 696	240	0·761
25	0·574	75	0·701	250	0·763
26	0·581	80	0·706	260	0·764
27	0·586	85	0·710	270	0·765
28	0·592	90	0·715	280	0·766
29	0·597	95	0·718	290	0·767
30	0·602	100	0·721	300	0·768

In America, scales of values for C are used in various localities for both impermeable areas and permeable areas (the latter meaning areas of low impermeability). As before mentioned, the writer recommends that in Great Britain the scale for impermeable area only should be used for developed areas, garden and parkland being taken as completely permeable. For Great Britain there are, as yet, no reliable run-off coefficients for permeable areas.

In Table 9 is given a typical 'rational'-method calculation. In this all the physical conditions, gross area, paved area, ultimate impermeability and available hydraulic gradients, are exactly the same as those used in the typical Lloyd-Davies calculation given in Table 7.

TABLE 9. Typical 'rational'-method calculation

Area covered by roofs and pavements (hectares)	Time of concentration (minutes)	Run-off coefficient	Rainfall intensity (millimetres per hour)	Run-off (cubic metres per minute)	Length divided by fall	Required diameter of sewer (nominal millimetres)	Capacity (cubic metres per minute)	Velocity (metres per minute)	Length (metres)	Time of flow (minutes)
0·65	5 say 11	0·406	51·3	2·26	350	300	3·7	50	244	5
1·46	5+7 = 12	0·427	48·5	5·04	360	375	6·5	57	387	7
4·86	12+3 = 15	0·478	42·0	16·2	325	525	16·6	74	220	3
6·50	15+3 = 18	0·515	37·3	20·8	500	675	25·9	70	215	3

TABLE 10. *Run-off per roofed or paved hectare*

Time of concentration (minutes)	Run-off per paved hectare (cubic metres per minute)	Time of concentration (minutes)	Run-off per paved hectare (cubic metres per minute)	Time of concentration (minutes)	Run-off per paved hectare (cubic metres per minute)
11	3·48	26	2·83	55	1·99
12	3·46	27	2·79	60	1·90
13	3·43	28	2·75	65	1·82
14	3·39	29	2·71	70	1·75
15	3·35	30	2·67	75	1·68
16	3·30	32	2·60	80	1·62
17	3·25	34	2·53	85	1·56
18	3·20	36	2·47	90	1·51
19	3·16	38	2·40	95	1·46
20	3·10	40	2·34	100	1·42
21	3·06	42	2·28	110	1·34
22	3·01	44	2·24	120	1·27
23	2·97	46	2·19	130	1·21
24	2·91	48	2·14	140	1·16
25	2·88	50	2·10	150	1·11

TABLE 11. Simplified 'rational'-method calculation

Area covered by roofs and pavements (hectares)	Time of concentration (minutes)	Run-off per paved hectare (cubic metres per minute)	Run-off (cubic metres per minute)	Length divided by fall	Required diameter of sewer (nominal millimetres)	Capacity (cubic metres per minute)	Velocity (metres per minute)	Length (metres)	Time of flow (minutes)
0·65	5 say 11	3·48	2·26	350	300	3·7	50	244	5
1·46	5+7 = 12	3·46	5·06	360	375	6·5	57	387	7
4·86	12+3 = 15	3·35	16·8	325	525	16·6	74	220	3
6·50	15+3 = 18	3·20	20·8	500	675	25·9	70	215	3

It will be observed that, in this case, the two methods result in the same diameters of sewer being required. But, on the average, the differences between the Lloyd-Davies method and the rational method, used with the author's selection of rainfall and run-off coefficient scales, are that the 'rational' method results in some economy in the smaller sizes of sewer that account for by far the greatest part of expenditure on sewerage works, a greater factor of safety for the larger sewers, overload of which could cause serious flooding, and a generally more consistent relation between sewer size and rainfall and run-off statistics.

The calculation may be still more simplified by multiplying the values of 0·16, C and R together to give a scale of figures which, multiplied by the appropriate roofed and paved area A will give the run-off in cubic metres per minute (see Table 10). But when this is done the interesting discovery is made that the scale so produced can be represented with remarkable accuracy by a formula of similar type to the Ministry of Health formula but of far more accuracy. This formula reads

$$Q = \frac{215A}{t+50}, \qquad (11)$$

where Q = run-off in cubic metres per minute,
A = roofed and paved area in hectares,
t = time of concentration in minutes.

Between the values of 12 and 120 minutes, the maximum deviations from the run-offs as calculated from the run-off and rainfall factors given in Tables 6 and 8 are no more than plus $2\frac{3}{4}\%$ and minus $1\frac{1}{4}\%$.

Table 11 gives the same calculation as given in Tables 7 and 9 but using the above formula. This table took 13 minutes to calculate and 10 minutes to check; in spite of imperial flow tables having to be used, there being no metric flow tables available at the time.

Tangent Method

While it is now evident that methods for estimating the effect of irregular distribution of impermeable area are of much less value than was once believed, there are rare occasions when they may be useful. These are when times of concentration are long and the bunching of impermeable area so marked as to be obvious on inspection: in most other circumstances such methods are a waste of time. Of these methods, the one of most utility is also the simplest. This is D. Wearing Riley's tangent method [20], which is applied as follows.

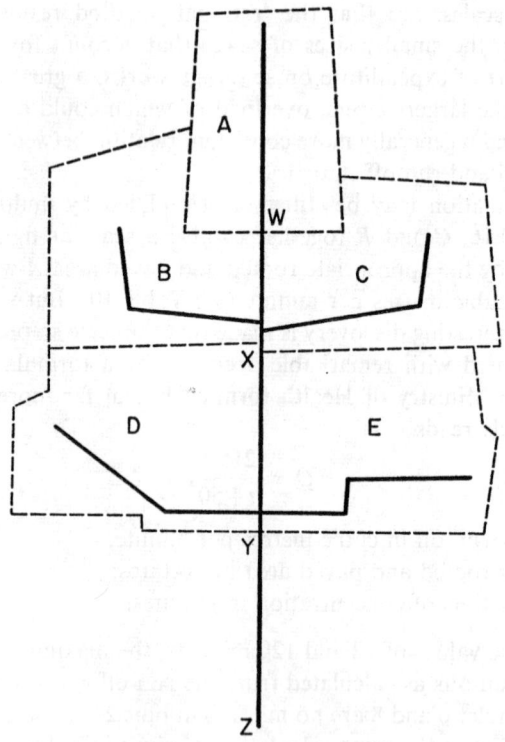

Fig. 1   Drainage area diagram

First, a time–area graph has to be prepared showing the impermeable area contributing to the point of outgo at various times after start of storm. Table 12 gives the data for the drainage area illustrated in Figure 1. To plot the time–area graph for the flow at point Y it is necessary to deal with each component area separately. Commencing with area D, the flow from the nearest part of the area is considered to arrive at Y when a storm starts and, after the time of concentration for this area (35 minutes), the whole of this area is draining to Y, and continues to do so for as long as rainfall continues. The curve for D is plotted as shown on the diagram (Fig. 2), rising

CALCULATING SEWER SIZES 43

to an assumed even rate from nil hectares impervious area at nil time to total hectares impervious area after the time of concentration, and then remaining constant.

TABLE 12. *Data for time–area graph*

Sewer	Drainage area	Roofed and paved area (hectares)	Time of concentration (minutes)
W to X	A	39·6	60
	B	39	40
	C	39	48
X to Y	A+B+C	117·6	60+4* = 64
	D	40·8	35
	E	30·6	30
Y to Z	A+B+C+D+E	189	64+18·4† = 82·4

* Time of flow from W to X, 4 minutes.
† Time of flow from X to Y, 18·4 minutes.

The curves for the other individual areas are plotted similarly, but areas B and C commence to discharge to the point Y after the time required to flow from X to Y, and A after the time to flow from W to Y. These curves are then summed to give the time–area graph.

The tangent method was originally based on the formula

$$R = \frac{X}{t+Y}, \qquad (12)$$

where R = intensity of rainfall,
       t = time of concentration,
       X, Y = appropriate constants as in the Ministry of Health formula.

As $Q = 0\cdot1\dot{6}\ C\ R\ A$, we can derive

$$Q = \frac{XA}{t+Y}$$

or, as in the formula already given,

$$Q = \frac{215A}{t+50},$$

44   SEWERS AND SEWAGE WORKS

Then, if from the point of maximum concavity of the time–area graph a distance is measured of minus 50 minutes and then a tangent is drawn to the time–area curve, a triangle is produced, the base of which measures $t+50$ and the height of which is the appropriate value of A (see Fig. 2). It will be observed that the value of Q varies as the tangent of the angle ϕ and therefore the greater this angle the greater the run-off.

In Figure 2, two trials have been made and it is found that the greatest value of ϕ is given when a distance of minus 50 minutes is measured from the point 0 minutes 0 hectares.

In this example the total time of concentration is 82·4 minutes and the total roofed and paved area 189 hectares. Thus, by the formula, the run-off is 307 cubic metres per minute. But the tangent method gives a greater run-off for part only of the area, i.e. that part which has a time of concentration of 58·4 minutes and an area of 165 hectares, giving a run-off of 327 cubic metres per minute or about $6\frac{1}{2}\%$ more than the run-off for the whole area.

The example given in Figure 2 was based on one published in 1939 and used to illustrate how the tangent method and similar methods could give greater run-offs than as calculated by the Lloyd-Davies method. At that time this example showed that the tangent method, used with the Ministry of Health rainfall curve, gave a run-off 14% more for the part area than for the whole area, while the Ormsby and Hart [11, 19] method gave 38% more. The recently acquired knowledge that run-off factor changes with duration of rainfall, has shown that these methods were giving fallacious values for run-off and also that the tangent method now rarely needs to be used.

Flat-rate calculations

When the time of concentration is less than about 11 minutes, the run-off coefficient decreases with reduction of time at a greater rate than the rainfall factor increases. Then the rule that the storm of duration equal to the time of concentration gives the greatest run-off no longer applies, and it is the 11-minute storm which gives the greatest run-off. For this reason, sewers serving all areas having times of concentration of 11 minutes and under should be designed on a flat rate of 51·3 millimetres of rainfall per hour which, at a run-off coefficient of 0·406, gives a run-off of 3·48 cubic metres per minute per roofed or paved hectare. This will apply to the majority

CALCULATING SEWER SIZES

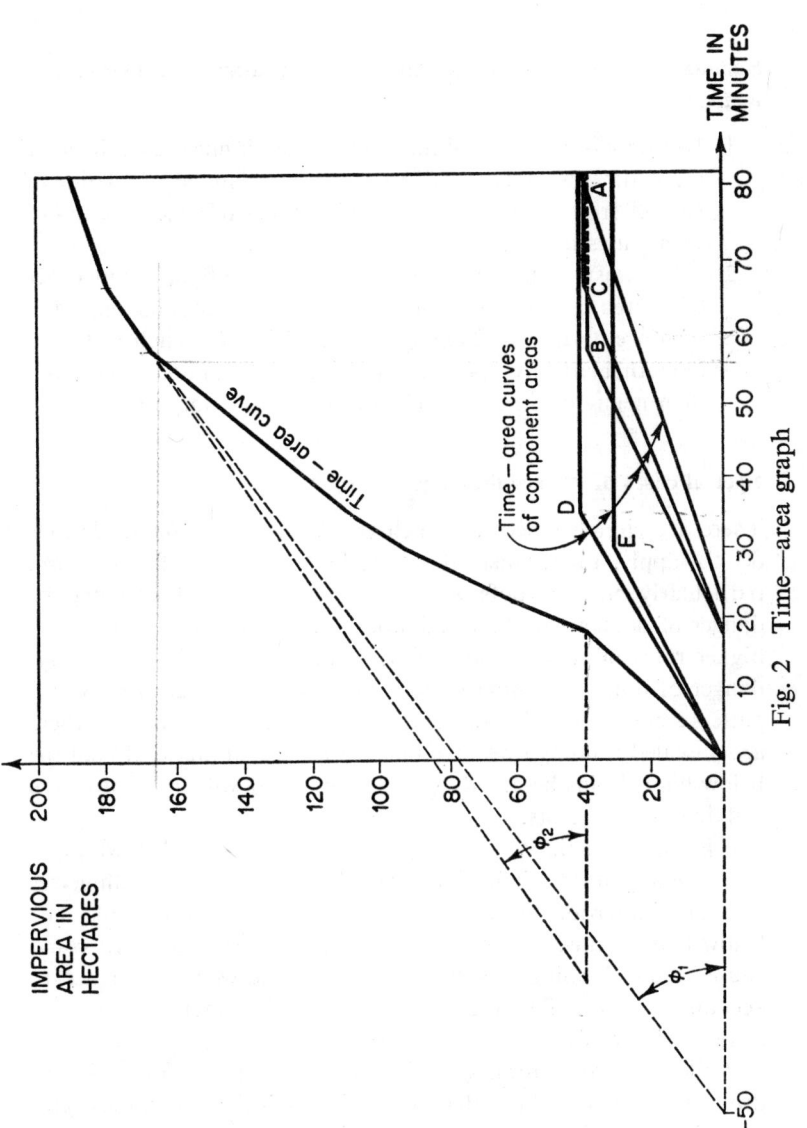

Fig. 2 Time—area graph

of rainfall run-off calculations, for most of the drains and sewers to be designed are those of small diameter and serving small areas.

Summary of conclusions on time-of-concentration and hydrograph methods

1. Not only is it a waste of time but, in fact, it must result in error if any time-of-concentration method is used in the calculation of run-off from drainage areas having times of concentration of 11 minutes or less.
2. It is a waste of time to apply the tangent method, or any other time–area graph or hydrograph method, to areas having times of concentration of 20 minutes or less, because such methods, if mathematically and statistically sound, cannot show more than negligible increases of flow in these circumstances.

Special cases of surface drainage

There are circumstances in which the foregoing recommendations do not apply. First, where there are large extents of factory roof, particularly pitched roofs with valley gutters, and there could be danger of damage to plant and stock should flooding occur, much higher rates of rainfall have to be assumed. There is a particular danger of indoor flooding when rainwater down-pipes from valley gutters connect to surface-water drains laid under the floors and there are unsealed rodding eyes or joints in the vertical pipes. Should the below-floor drains be greatly surcharged, water will spill out of the rodding eyes or joints.

It is usual practice to design rainwater guttering and down-pipes to accommodate a rainfall intensity of not less than 75 millimetres per hour from roof surfaces assumed to be 100% impermeable. The below-floor drains, or any other drains, surcharge of which could cause indoor flooding or serious damage, could well be designed to accommodate this flow (12·5 cubic metres per minute per hectare) instead of any of the figures given in Table 10.

While aerodrome drainage might be considered a form of land drainage involving quite different principles of calculation, the drainage of runways is most frequently designed in the same manner as surface-water sewerage because of the large paved runway areas and the rapid run-off from specially drained grassland.

The difference between the drainage of aerodrome grassland and agricultural drainage is that field drains usually follow the slope of the land, whereas aerodrome drains are laid in rubble-filled trenches parallel to the contours and fairly close together so that the water quickly finds its way from the surface to the drains.

Soakaway systems

Where the subsoil is permeable, surface water can be discharged to soakaways. In fact it *should* be so discharged, for in localities where there is no natural surface run-off, the provision of surface-water sewers will involve difficult man-made problems.

There is seldom any doubt that soakaways, and not surface-water sewers, should be provided in localities such as chalk and other limestone areas where there are no natural watercourses except perhaps an occasional small stream or winter-bourne rarely to be found in major valleys. If surface-water sewers were to be constructed in such places, they would have to be laid for considerable distances to connect to some watercourse quite incapable of taking the flow.

In some areas, such as large tracts of gravel or similar porous subsoil, it may not be clear as to whether soakaways should be used, although they may work satisfactorily. These places have to be treated on their own merits.

In any one locality the performance of individual soakaways can be very varied. For example, in chalk areas some of the soakaways may be capable of taking the discharge of a fire hydrant without showing any sign of filling with water, while others may be almost useless. Consequently it is usually advisable to connect soakaways together with small-diameter drains (to which road gullies may be connected) so that if one soakaway is unsatisfactory it can spill over to the next. An extra-large soakaway should be placed at the bottom end of each line.

Soakaways are large chambers about 10 metres deep and 3 metres diameter. They should *not* be filled with rubble for it is storage capacity which is important. Fair practice is to make each soakaway capable of storing the equivalent of about 15 millimetres of rainfall over the paved or roofed area which it serves. Any part of the chamber capacity below the standing ground-water level must be excluded from the calculation.

3

Sewage Pumping

In the design of sewage-pumping stations, more precautions have to be taken than in the design of similar installations for water supply etc., for the pumping stations used in connexion with sewerage and sewage-treatment works mostly have to deal with liquids containing solids such as crude-sewage organics, screenings, detritus of various sizes, sewage sludge and even pieces of timber, butcher's waste and various large objects accidentally or illicitly discharged to the sewers.

Ability to pump fluids containing these various solids without mechanical breakdown or the development of nuisance calls for not only the right selection of pumps but also the proper design of incoming sewer, suction well, valves, delivery pipework and rising mains.

The types of pumps most frequently used for pumping sewage and similar liquids are volute centrifugal pumps for delivering against heads of 6 to 36 metres, axial-flow pumps particularly used for delivering large quantities against low heads with little or no suction head, and mixed-flow pumps, the characteristics of which came generally between those of volute and axial-flow pumps.

For small sewage-pumping stations, in particular those of the smallest size delivering to rising mains of 100 millimetres diameter (the minimum diameter recommended for crude sewage), unchokable volute pumps are most frequently used. The secret of the design of an unchokable pump is to have impeller blade or blades with no leading edge over which fibrous material can wrap and build up to cause chokage. Suitable impellers are the Wallwin single-blade open-type impeller which has one blade in the form of an S with both ends at the periphery, or similar impellers having two S-shaped continuous blades parallel one with the other. In each case the eye of the pump enters clear of any part of the impeller.

Multi-blade pumps as used for water, and having a number of

blades each of which has a leading edge starting near the eye of the pump, are almost certain to choke if used for crude sewage unless of very large size.

Pumps for sewage should never have inlets of less than 75 millimetres diameter and are still better if the inlets are 100 millimetres diameter, for then the inlet is as large as the smallest drain delivering to the sewer and should be able to take the largest solids that have entered the sewer by legitimate means.

Axial-flow and mixed-flow pumps can be very useful for dealing with crude sewage (except in the smaller sizes) provided that the suction and delivery heads are limited to the capabilities of the impeller. They are particularly useful for delivering returned activated sludge or for the recirculation of effluent in recirculation percolating-filter schemes.

Turbine pumps are generally not fit for any sewerage or sewage-treatment purposes.

At one time three-throw reciprocating pumps, usually of the plunger type, were largely used for sewage pumping or for dealing with thick sewage sludges. They are now considerd too costly and, with the improvement of centrifugal pumps, unnecessary. Reciprocating pumps for sewage have to be protected by screens, which is an added disadvantage. They have one advantage which at times could decide on their use: except at very low heads, the power they require varies almost directly as the head. The efficiency is good at all except low heads. A reciprocating pump could well be a good choice for delivering sewage to an exceptional head of more than 36 metres.

In addition to ordinary mechanical pumps of centrifugal or reciprocating type, there are air-operated devices such as sewage ejectors and low-lift air lifts which can be very useful.

Centrifugal pump characteristics

Unlike reciprocating pumps, all types of centrifugal pumps deliver less quantity as the manometric head is increased and vice versa, according to the characteristic curve of the pump. Also, the quantity delivered varies directly as the speed of rotation of the impeller, and the manometric head can be increased as the square of the speed of rotation. The power demand depends on the quantity delivered multiplied by the manometric head and is, of course, affected by the efficiency. The last is at a maximum at some intermediate point on the characteristic curve which relates manometric head to quantity delivered.

The shape of the characteristic curve varies with the type of centrifugal pump, and the curve of power input varies also. A volute pump requires an increase of power with increase of delivery and reduction of manometric head, and if the head is reduced beyond the point of maximum efficiency, the power demand may increase so much as to overload the motor. The reverse is true of an axial-flow pump, for which the power input rises on increase of head and reduction of delivery. A mixed-flow pump has intermediate characteristics usually tending to resemble those of the axial-flow pump more than those of the volute pump.

Specific speed

Specific speed is a term used to classify axial-flow, mixed-flow and volute centrifugal pumps on the basis of their performance and dimensional proportions, regardless of their size of speed of rotation. It expresses the speed in revolutions per minute of an imaginary pump geometrically similar to the actual pump under consideration and capable of raising 75 kilogrammes of water per second to a height of 1 metre (1 metric horse-power, cheval-vapeur or Pferdestarke). The metric formula for specific speed reads

$$N_s = 3.65nQ^{0.5}/H^{0.75}, \qquad (13)$$

where N_s = specific speed,
n = revolutions per minute,
Q = delivery in cubic metres per second,
H = manometric head in metres at maximum efficiency.

Specific speed is calculated in America in different units from those that were used in Great Britain prior to the adoption of the metric system. Thus, when comparing specific speeds calculated in different countries, it is necessary to know the units and constants that were used, for the value of N_s found for a pump in one country will vary from that found for the same pump elsewhere.

At the time of writing no metrication committee recommendation has been made on specific speed: the above formula was provided to the author by a well-known international firm of pump manufacturers.

Types of installation

The majority of pump sets used for pumping sewage or the various sewage-works liquids are of the vertical-spindle dry-well type. This is by no means the least expensive arrangement but it has a number of marked advantages. First, the electric motor or the prime mover can be placed above ground level or such other level as is necessary to ensure that it cannot be damaged by flooding with sewage or storm water. Second, the pump itself is placed in a below-ground chamber or 'dry well' easily accessible for maintenance purposes. Thus, by being set at a level below that of the incoming sewage, the pump is automatically primed, thereby doing away with the necessity for priming gear. The incoming sewage falls into a wet well that is isolated from the rest of the pumping station by watertight structures. From this the suction pipes of the various pumps draw.

Other types of pump, such as vertical-spindle submersible pumps and horizontal-spindle pumps, are used for sewage pumping. Submersible pumps are inexpensive and, in the smaller sizes, are useful at sewage works serving isolated buildings. But in large sizes they are not desirable because they are difficult to lift out for maintenance and cause the mechanics to come in contact with sewage, with risk of infection. Horizontal-spindle pumps usually require that the motor shall be placed at a low level with risk of its being flooded, and it is surprising how often this happens.

Cavitation

If a pump is called upon to suck against a head greater than the atmospheric pressure, a partial vacuum is formed. If the negative pressure in the suction pipework is that at which the liquid will vaporize, the effect is similar. Thus, at ordinary day temperatures it is impracticable to lift by suction against a manometric head of more than about 10 metres. In practice it is usual not to allow for a suction head greater than about 7 metres for clean water or one-half that amount for sewage sludge.

Apart from this limitation, axial-flow pumps and, to a lesser extent, mixed-flow pumps, are very liable to cavitation behind the blades of the impeller and this causes pitting and honeycombing of the impellers. (The presence of cavitation is made evident by a rattling noise.) Accordingly it is usual so to arrange axial-flow pumps that there is little or no suction lift, and to set a limit on the suction lift of mixed-flow pumps.

Planning the layout of a pumping station

The best arrangement for a pumping station dealing with one type of flow only, say dry-weather flow, is to have a number of pumps all identical in size, direction of rotation, etc., equally spaced in a straight line. If there are pumps serving two purposes, such as dry-weather flow and storm-water pumping, they may still be all in one straight line but there will now be two sizes of pump and two spacings apart according to those sizes. Alternatively there may be two parallel lines of pumps. The latter arrangement, while effecting some economy in the cost of the building, may more than double the cost of the gantry crane or cranes.

A very common practice is to have several sizes of pump, a small pump to cut in on small flows, a larger one to cut in on larger flows the small one then cutting out, a still larger one to cut in on a still greater flow and so on. This arrangment has a number of disadvantages: it calls for a greater number of stand-by pumps and a greater number of spare parts, and introduces the possibility of complete breakdown of the station should the electrical co-ordinating device fail.

The vast majority of modern sewage-pumping stations are automatic electric stations which can be left unattended. For these to be completely reliable, it is necessary that the failure of any one pump or part of the electric gear shall not cause the station to fail as a whole. To ensure reliable operation, it is best to have two separate electricity supplies not liable to fail at the same time, and to connect one supply to half the total number of pumps and the other supply to the remainder. Furthermore, each pump should be considered a separate entity started and stopped by its own floats, electrodes or other control gear, and not interconnected with any other pump by an electric co-ordinating device. Complication in an automatic electric pumping station is always liable to lead to trouble, and additional complication by electronic warning devices or automatic switch-over from one power supply to another is no protection.

In every sewage-pumping station there should be no fewer than one stand-by pump, and in any important sewage-pumping station, more than one stand-by pump is desirable. No pump should be so small that it does not deliver enough to keep the rising main clean. On the other hand, the number of pumps should not be small except in very small stations having one duty and one stand-by pump. It is a common misconception that an installation consisting of several small pumps is much more costly than one having a few large pumps. The fact is that while the pumps and motors themselves in a station made up of many small pumps may be more expensive than the pumps and motors in a station provided with a few large pumps, the extra cost is offset by the very high costs of the sluice valves and reflux valves in the station containing the large pumps. And so it is that, between very wide limits of pumping station size and pump size, the cost of installation of pumping plant is in almost direct relation to the total pumping capacity in cubic metres per minute of the installed plant.

Dimensions for drawing purposes

It is necessary to design the pumping station in more than outline before tenders are obtained and pump manufacturers' detailed drawings received. To be able to do this the designer must have sufficient knowledge of the normal or average proportions of pumps, valves and pipework.

Pumps by different makers vary in their dimensions but there are some similarities and, for determing proportions near enough for drawing purposes, the designer can use the formula

$$D = KQ^{0.5}/H^{0.25}, \qquad (14)$$

where D = diameter of suction branch in millimetres,
 Q = delivery of volute centrifugal or mixed-flow pump in cubic metres per minute,
 H = total manometric head in metres,
 K = a constant.

The value of K can usually be taken as 156 but, of the makes of pump studied when this formula was derived, the maximum value was 180 and the minimum the somewhat undesirable figure of 130.

Having found the nominal internal diameter of the suction branch one can determine the other dimensions of a volute or axial-flow centrifugal pump near enough for practical purposes by equations 15 to 18.

Distance back from face of flange of volute pump delivery to a plane passing through axis of pump

$$= D + 280 \text{ millimetres} \qquad (15)$$

Distance from axis of pump to axis of delivery

$$= D + 178 \text{ millimetres} \qquad (16)$$

Distance from face of flange of suction to a plane passing through axis of delivery

$$= D + 100 \text{ millimetres} \qquad (17)$$

Minimum distance between axes of vertical spindle pumps

$$= 2 \cdot 5D + 1280 \text{ millimetres} \qquad (18)$$

Suction and delivery pipework and valves

The appropriate diameters of suction and delivery pipes are important and should be specified, for otherwise the competitive tenderers may reduce the sizes of the pipes and valves more than is desirable. These pipes should be of the economic sizes which, for flanged cast-iron pipes erected indoors, are approximately as given in Table 13.

It may be convenient to make the diameters greater than given in the table but this would add to costs and occupy more space. Generally they should not be smaller, and in no case should a suction or delivery pipe be smaller than the suction or delivery branch of the pump to which it connects.

While pumping-station pipework is often made to non-standard proportions by specialist firms, it is best to design according to the dimensions of cast-iron pipe fittings given in the British Standard Specification for Flanged Pipes. Incidentally, all indoor pipework should be flanged because of the danger of joints being blown open by pressure. But it should be remembered that a straight line of flanged pipe built into walls at both ends cannot be taken down and re-erected: there must be a bend in the line or one or more diagonal flanges to make this possible.

Sluice valves suitable for sewage and reflux valves are not standardized and may vary very much in their proportions. But the following rules are near enough for drawing purposes:

SEWAGE PUMPING   55

Distances between flanges of sluice valves

$$= 30 D^{0.5} \text{ millimetres} \qquad (19)$$

Height of sluice valve measured from axis of pipe to cap of spindle

$$= 2D + 382 \text{ millimetres} \qquad (20)$$

Distance between flanges of reflux valves

$$= 2D + 152 \text{ millimetres} \qquad (21)$$

(In large sizes this last rule gives excessive lengths because the type of reflux valve changes, but the rule still proves useful.)

TABLE 13. *Economic diameters of pipework and valves*

Discharge of one pump (cubic metres per minute)	Economic diameter of suction and delivery pipework and valves (nominal millimetres)	Approximate economic velocity (metres per second)
0·40	100	0·823
0·68	125	0·885
1·08	150	0·975
1·53	175	1·04
2·21	200	1·13
2·97	225	1·22
3·85	250	1·28
6·09	300	1·40
10·65	375	1·52
16·8	450	1·71
24·7	525	1·83
34·4	600	1·95
46·2	675	2·07
60·2	750	2·22
76·3	825	2·32
95·0	900	2·41

The suction pipes usually draw upwards from the bottom of the suction well, bends being provided for the pipes to turn horizontally through the wall between the suction well and the dry well. Small-diameter suctions for sewage should not have bellmouth inlets, for this encourages choking. Rectangular slot-shaped ends are sometimes used but plain ends are most common.

An alternative arrangement for small installations is for the suctions to draw downwards from the bottom of the well. In either case the lowest level of sewage in the well should be near the bottom but not lower than a distance equal to the velocity head above the inlet of the suction so that vortex action will not cause air to be drawn in.

The suction pipe passes through the wall with a suitable puddle flange to prevent leakage and with the flanged joints 150 millimetres clear of the wall to give access to the bolts. A sluice valve is fixed between the wall and the pump to isolate the wet well from the dry well.

On the delivery side of the pump there is a necessary taper pipe, a reflux valve and a sluice valve to isolate the pump and reflux valve from the rising main. The deliveries of the several pumps connect laterally, not vertically, into the rising main.

In all except small installations, it is desirable to have two rising mains for two reasons. First, the provision of two rising mains makes it possible to keep one in operation while the other is under maintenance. Second, the velocity in the rising main is usually limited between the self-cleansing velocity of 46 metres per minute and a preferred maximum of 146 metres per minute, and if it is desirable to have several pumps, two rising mains make twice the number of pumps possible, or a wider range of maximum to minimum flow.

Sluice valves for sewage should *not* be British Standard valves as used for waterworks purposes but should have means of access for removal of any heavy solids that might prevent the gate from closing, and those inside pumping stations or dry chambers should have external, not internal, screws. Handwheels should be placed above the highest possible flood level. Reflux valves for sewage should be a single-gate type, for multi-gate valves can choke with fibrous material. If the valves have external arms connected to the flaps, these must be properly adjusted and weighted, or damage by slamming is very likely to occur. Reflux valves should always be placed in horizontal pipework: while valves are designed for use in vertical pipework, these are not serviceable for sewage or sludge because solids gravitate down the vertical pipe to settle behind the gate, causing it to stick open with danger of slamming and damage.

Owing to the risk of damage by reflux valves slamming, it is always desirable to make some provision to absorb shock. One method is to place an air vessel on the top side of the rising main as near the valves as possible: this must have means for recharging with air. Another is to place near the station a stand pipe connected to the rising main and capable of discharging back to the suction well if the pressure in the main greatly exceeds the maximum delivery pressure envisaged in normal operation: this must have an air inlet at the top so that it will not siphon back the contents of the rising main.

In addition to the main suction and delivery pipework, small-diameter suctions are useful for removing from the dry well any infiltration or any liquid spilt upon the floor during maintenance. This is in addition to any automatic cellar-emptying pump that may have been installed. The handwheels of these valves must be carried to above flood level. In small pumping installations it is good to have means of blowing back from the delivery of one pump to the suction of another to clear any chokage. A washout connexion from the rising main to the suction well is practicable if either the suction well is large enough to take the entire contents of the rising main or there are two rising mains, one of which can be in operation while the other is being emptied, but not otherwise.

In the design of other than small pumping stations, the practical use of valves and penstocks must be considered. No good purpose can be served by putting in a valve that is to be hand operated in an emergency if it is going to take half an hour to close, and therefore large valves, except those that are seldom used and never need to be opened or closed quickly must be power operated. Similarly, no valve should have a handwheel so small or gearing of such ratio that it cannot be turned by hand.

The design should allow sufficient space for handwheels of adequate size. The approximate diameter of ungeared handwheels is about the same as the distance between the flanges of a flanged sluice valve, and can be determined by the formula

$$d = 30D^{0.5}, \qquad (22)$$

where d = diameter of handwheel in millimetres,
D = internal diameter of valve in millimetres.

Large valves need to be geared, the usual ratios being 2:1, 3:1, 4:1 and, in the case of worm gears, about 20:1.

A number of tests made on a wide range of sizes of valve showed that the opening or closing by hand of geared or ungeared valves or penstocks took about 54 minutes per square metre of waterway, or

$$t = D^2/23600, \qquad (23)$$

where t = time in minutes,
 D = internal diameter in millimetres.

The direct pull on spindle required to open a valve or penstock which is lifted from above is given approximately by the formula

$$P = 350A\,(H+1), \qquad (24)$$

where P = pull in kilograms (including friction and weight of door),
 A = area of waterway in square metres,
 H = unbalanced head in metres.

Automatic operation

For as long as can be remembered it has been usual to build completely automatic electric sewage-pumping stations wherever electricity was available, but it is only in comparatively recent years that the largest of sewage-pumping stations have been made entirely automatic.

The difficulties that have been experienced in sewage-pumping station design have mainly been due to the solids in the sewage. Early practice was to construct unduly large suction wells which acted as unwanted settlement tanks in which sludge and detritus collected at the bottom and scum at the surface. These wells had to be cleaned out at intervals and this was not a pleasant operation. Also subsidence of detritus has been known to choke the suctions, putting the station out of action. To overcome these difficulties, the suction well needs to be small enough for the turbulence of the incoming flow to prevent settlement and scum separation, and it must be so designed that it is regularly emptied and solids do not accumulate on the sloping floors. In former times it was usual to install screens, sometimes manually cleansed, to protect the pumps. Now screens are used only in the larger stations where the pumps are of too great a size to be of the unchokable type and the screens must be automatically controlled.

Pumps are automatically started and stopped by float switches, electrodes, pneumatic gear or some similar device, floats and electrodes being most common in Great Britain. One arrangement is to have a float in a vertical tube, the purpose of which is to prevent it from being washed about by turbulence. The float is suspended on a non-corrodable cable that passes round a drum and operates a float switch so that a pump is started when the float reaches a predetermined high level and stops when it reaches a low level. A more favoured arrangement is to have a float, with a hole through the middle, passing freely up and down a rod between stops the positions of which can be adjusted according to desired top and bottom water levels. The floats are made of non-ferrous metal or ceramic material, and should be so designed that they float half submerged.

In the arrangement of float gear care should be taken so that the float will not be able to rest on a bank of sludge where it may stick and that it will not be influenced by currents or turbulence. If a float rod or wire passes from a wet well to the motor room or dry well, care must be taken that this does not make possible the passage of foul air to the motor room or dry well, or flooding of the dry well.

Electrodes for pump operation consist of two rods suspended from above water level. One causes a pump to start when the water rises to touch it, the other makes a pump stop when the water falls so as to break contact. Low-voltage current is supplied to the rods and a system of delicate relays actuates the starting gear. When designing the arrangement of electrodes it must be remembered that lateral support is not practicable and therefore the electrodes must be so placed that turbulence does not make them swing about. They must also be readily accessible for cleaning and, if necessary, removal. Each pump must have its own pair of electrodes.

There are other special devices obtainable, including pneumatic pressure-operated starting and stopping switches, a hinged float containing a magnet which operates a switch in a sealed unit, and a switch operated by a beam from radio-active material.

Dimensions of suction well

The suction well is made small to avoid separation of solids, but it must not be too small or the pumps will cut in and out so frequently as to overheat the starting gear. Standard rheostat starters are in three classes, ordinary duty, which can be started twice per hour, intermittent duty, which can be started fifteen times per hour, and frequent duty, which can be started forty times per hour.

A practice which has always proved adequate is to specify frequent-duty starters but to size the suction well so that no pump starts more frequently than fifteen times an hour: this gives a good factor of safety. A pump starts most frequently when the rate of flow is equal to half its delivery. If the pump is of such a capacity that it can empty the well from cut-in to cut-out level in 1 minute when there is no flow then, when there is inflow at half the pumping rate, the well will be pumped out in 2 minutes and, when the pump has stopped, refill in another 2 minutes, giving a start to restart cycle of 4 minutes. Thus, if a pump is not to start more frequently than fifteen times per hour, the capacity between its cut-in levels and cut-out levels must be the equivalent of 1 minute's pumping rate of that pump.

When there are several pumps all of the same capacity, all that is necessary is that the cut-in level of the second pump is a little higher than the cut-in level of the first pump, that the cut-out level is the same amount higher, and so on, according to the number of pumps installed. With this arrangement, if the first pump is beaten by the flow, the sewage level rises to cut in the second pump. If this beats the flow, the well is pumped out and the second pump stops, leaving the first pump running, but if the flow beats the second pump, a third pump cuts in. Should any pump set fail to work, the next automatically cuts in, and so the operating of the pumping station is not upset.

It is usual to have means by which the floats or electrodes can be changed round so that any pump can be made the first to cut in. This arrangement is always desirable, but it is not good practice to keep changing the order of the first pump to cut in, as this may lead to all pumps wearing out and therefore coming on maintenance at the same time instead of in turn.

The distance between the cut-in levels of the various pumps should never be less than 150 millimetres, because some float switches do not move until the float has been submerged 100 millimetres and also because turbulence can cause two pumps to cut in together if there is insufficient separation. Allowance has also to be made for the time taken for the pump to attain working speed; centrifugal pumps do not deliver at all until the peripheral speed is sufficient to overcome the dead lift. Table 14 gives the approximate starting times in seconds for various sizes of pump.

Pumping directly from the sewer

When sewers are of very large diameter it is possible to pump directly from them without the construction of any *ad hoc* suction well, provided that proper precautions are taken. The storage capacity

TABLE 14. *Starting times of electric pumps (seconds)*

Type of pump	37 kW	112 kW	223 kW	373 kW	746 kW
Centrifugal	20	30	40	60	90
Reciprocating	20	40	60	60	90

between cut-out and cut-in levels is the wedge of water between the level at the bottom end of the sewer when it is flowing partially full and freely discharging to the pump (it is safest to neglect drawdown) and the level in the sewer when it is backed up by the pumps being beaten. In the latter case it can be assumed that the sewage lies level from the bottom end to the point where it meets the surface of flow parallel with the invert. This gives a wedge-shaped body of water the capacity of which can be calculated. Table 15 can be used for estimating the depth occupied in a sewer running partly full.

TABLE 15. *Proportional depths and areas and probable proportional velocities and discharges of circular pipes and culverts flowing partly full*

Proportional depth	Proportional area	Proportional velocity	Proportional discharge
0·1	0·0520	0·3519	0·018
0·2	0·1424	0·5267	0·075
0·3	0·2523	0·6547	0·165
0·4	0·3735	0·7523	0·281
0·5	0·5000	0·8300	0·415
0·6	0·6265	0·8859	0·555
0·7	0·7477	0·9228	0·690
0·8	0·8576	0·9381	0·805
0·9	0·9480	0·9261	0·878
1·0	1·0000	1·0000	1·000

In making calculations of this kind care must be taken that all cut-in levels are well below the crown of the sewer, otherwise surge may cause sudden excessive pressures on the suction side of the pumps. On the other hand, experiments showed that it was not possible to produce surge in a sewer that never ran full.

Capacities of pumping stations

A pumping station for dry-weather flow should be designed to take the maximum rate of flow of four or six times dry-weather flow with its duty pumps, and to have an additional pump or pumps as stand-by but not arranged to come into operation automatically. If there are two duty pumps and one stand-by, there should be two float or electrode installations only, not three.

The circumstances are different in the case of a storm-water pumping station for, as was described in Chapter 2, storm-water sewers are designed on the assumption that surcharge and other factors will make them capable, when necessary, of dealing with greater flows than those estimated on the basis of the design storm. A pumping station has not this flexibility: if the flow beats the maximum capacity of the pumps, flooding will occur. It is therefore desirable to make storm-water pumping stations capable of delivering more than the calculated discharge of the storm-water sewers and, as an additional factor of safety, to have the float or electrode gear so arranged that all stand-by pumps will come into operation except any that may have been taken down for maintenance purposes.

It is often practicable to store storm water in the suction well so that the rate of flow into the pumping station can exceed the pumping rate during the storm. This can either make possible the installation of smaller pumping capacity and rising main than would be required according to the rainfall run-off formula, or provide an additional factor of safety. For the former purpose, the storage capacity must be above the highest cut-in level of any pump but below the crown level of the incoming sewer. But if one desires to calculate the effect of such storage as may be available before flooding occurs, the storage capacity may include all that is available below the flood level at the pumping station. The relation of storage capacity to the frequency with which it will be fully occupied can be calculated by the formula

$$C = 62 A p^{1.5} N^{0.5}/P^{0.5}, \qquad (25)$$

where C = storage capacity in cubic metres,
Ap = impermeable area in hectares,
N = number of years between occurrences of storm,
P = pumping rate in cubic metres per minute.

The practice of some engineers of arranging for spill-over of untreated sewage from a soil-sewage pumping station to a watercourse is contrary to Sections 30 and 31 of the Public Health Act, 1936 and should not be permitted.

Power calculations

The manometric head against which a pump has to deliver, so called because it is the total head as measured by pressure gauge, is made up of the dead lift from the highest level of sewage in the suction well to the crown on the highest part of the rising main plus the various losses at entry, through the valves, pipework and rising main as far as the point to which the maximum lift has to be made. The loss at entry, known as velocity head, is the head required to accelerate the water from stationary to the velocity in the pipework (and which may be partially recovered). It can be calculated by the formula

$$H = V^2/2g, \qquad (26)$$

where H = velocity head in metres,
 V = velocity in metres per second,
 g = acceleration under gravity.

The value of g varies slightly with latitude, the international metric value being 9·80665 metres per second per second.

TABLE 16. *Approximate loss of head through fittings*

Type of fitting	Equivalent length of straight pipe of equal diameter
Sluice valve	6 × pipe diameter
Reflux valve	50 × pipe diameter
Bend (radius equal to 3–5 diameters)	14 × pipe diameter
Round elbows	30 × pipe diameter
Sharp elbows and T-connexions	90 × pipe diameter

The loss of head through sluice valves, bends, junctions, etc., varies with diameter, but the rule of thumb as given in Table 16 is sufficiently accurate for practical purposes. Table 17 gives rather more accurate figures for reflux valves for velocities up to 2·4 metres

TABLE 17. *Loss of head through reflux valves*

Velocity in metres per second	Loss of head in metres
0·6	0·110
0·9	0·117
1·2	0·128
1·5	0·143
1·8	0·158
2·1	0·176
2·4	0·206

per second. In this range, which is the practical range of velocities, the flap of the valve is not fully open and resists the flow. The loss of head through the pump is included in the pump efficiency: that through the pipework of the pumping station and through the rising main can be calculated by pipe-flow formula or tables.

Power requirement is calculated as follows. Assuming that a cubic metre of water or sewage weighs 1000 kilogrammes, the pressure on the inside of the pipework is 0·1 kilogramme per square centimetre per metre of head. On the same assumption, and when the total manometric head has been found by adding together the dead lift and the various head losses in metres, the required power of the motor in kilowatts can be found by the formula

$$\text{kW} = \frac{QH}{6 \cdot 1183} \times \frac{100}{\text{pump efficiency (\%)}}, \qquad (27)$$

where kW = *mechanical* load on motor in kilowatts,
Q = cubic metres per minute,
H = manometric head in metres.

This value, multiplied by 100 and divided by percentage motor efficiency, gives the electricity demand in kilowatts which, multiplied by 100 and divided by percentage power factor, gives kilovolt amps on which kVA charges are made. Approximate efficiencies and power factors of motors are given in Table 18.

The main charge for electricity is on the consumption in kilowatt hours, but frequently there are additional charges for kilovolt amps

TABLE 18. *Efficiencies and power factors of electric motors (experimental values)*

Kilowatts at 100% motor efficiency	Efficiency (%)			Power factor (%)		
	Full load	¾-load	½-load	Full load	¾-load	½-load
1·5	74	73	70	80	75	65
2·25	78	78	76	81	76	66
3·0	81	81	79	82	78	69
7·5	85	85	83	85	81	72
15·0	88	89	88	89	87	80
30·0	89	89	87	90	88	82
75·0	91	92	91	88	85	79
150·0	93	93½	93	89	87	81

as based on the size of the installation and sometimes on maximum peak demand as recorded on the meters.

To find the annual running cost of a pumping station, the number of pump hours per day required to deliver the average daily flow (which is not the dry-weather flow) multiplied by the kilowatts required by the number of pumps running gives kilowatt hours per day. This multiplied by 365¼ gives kilowatt hours per year. To this should be added the annual kVA charge and the maintenance costs of the pumping station, including all labour and materials. The last (at the time of writing, 1969) can be estimated by the formula

$$\text{Annual maintenance cost} = £50 \times Q^{\frac{1}{3}} \qquad (28)$$

where Q = normal duty of the station in cubic metres per day.

TABLE 19. *Example calculation of optimum diameter of rising main*

Diameter of pipe (millimetres)	Electricity charge (£'s)	Repayment of loan (£'s)	Total annual cost (£'s)
200	772	77	849
225	533	86	619
250	412	92	504
300	316	112	428
350	278	133	**411**
375	272	140	412
450	261	171	432

TABLE 20. *Repayment of loans (£'s annual repayment, including capital and interest, per £100 of loan)*

(%) interest	Period of loan		
	15 years (machinery)	30 years (works)	60 years (land)
2	7·7825	4·4650	2·8768
2⅛	7·8555	4·5421	2·9645
2¼	7·9289	4·6199	3·0535
2⅜	8·0026	4·6985	3·1438
2½	8·0767	4·7777	3·2353
2⅝	8·1511	4·8577	3·3281
2¾	8·2259	4·9384	3·4220
2⅞	8·3011	5·0198	3·5171
3	8·3767	5·1019	3·6133
3⅛	8·4526	5·1847	3·7106
3¼	8·5289	5·2682	3·8090
3⅜	8·6055	5·3523	3·9084
3½	8·6825	5·4371	4·0088
3⅝	8·7599	5·5226	4·1103
3¾	8·8376	5·6088	4·2127
3⅞	8·9157	5·6956	4·3160
4	8·9941	5·7830	4·4202
4⅛	9·0729	5·8711	4·5253
4¼	9·1520	5·9598	4·6312
4⅜	9·2315	6·0492	4·7379
4½	9·3114	6·1392	4·8454
4⅝	9·3916	6·2298	4·9537
4¾	9·4721	6·3210	5·0627
4⅞	9·5530	6·4128	5·1724
5	9·6342	6·5051	5·2828
5⅛	9·7158	6·5981	5·3939
5¼	9·7977	6·6917	5·5056
5⅜	9·8800	6·7858	5·6178
5½	9·9626	6·8805	5·7307
5⅝	10·0455	6·9758	5·8441
5¾	10·1288	7·0716	5·9581
5⅞	10·2124	7·1680	6·0726

TABLE 20 (*continued*)

(%) interest	Period of loan		
	15 years (machinery)	30 years (works)	60 years (land)
6	10·2963	7·2649	6·1876
6⅛	10·3805	7·3623	6·3030
6¼	10·4651	7·4603	6·4189
6⅜	10·5500	7·5588	6·5353
6½	10·6353	7·6577	6·6521
6⅝	10·7208	7·7572	6·7692
6¾	10·8067	7·8572	6·8868
6⅞	10·8929	7·9577	7·0047
7	10·9795	8·0586	7·1229
7⅛	11·0663	8·1601	7·2415
7¼	11·1535	8·2620	7·3604
7⅜	11·2409	8·3643	7·4796
7½	11·3287	8·4671	7·5991
7⅝	11·4168	8·5704	7·7189
7¾	11·5052	8·6741	7·8390
7⅞	11·5939	8·7782	7·9593
8	11·6830	8·8827	8·0798
8⅛	11·7723	8·9877	8·2006
8¼	11·8619	9·0931	8·3215
8⅜	11·9518	9·1989	8·4427
8½	12·0421	9·3051	8·5641
8⅝	12·1326	9·4116	8·6857
8¾	12·2234	9·5186	8·8074
8⅞	12·3145	9·6259	8·9293
9	12·4059	9·7336	9·0514
9⅛	12·4976	9·8417	9·1737
9¼	12·5896	9·9501	9·2960
9⅜	12·6818	10·0589	9·4185
9½	12·7744	10·1681	9·5412
9⅝	12·8672	10·2775	9·6640
9¾	12·9603	10·3873	9·7868
9⅞	13·0537	10·4975	9·9098

68 SEWERS AND SEWAGE WORKS

The above formula is based on cost records of automatic sewage-pumping stations which were not manned but visited by a travelling maintenance gang who had to deal with several pumping stations.

Sewage rising mains constructed of cast-iron pipes (Class B in the majority of instances), steel pipes or asbestos-cement pressure-pipes are most frequently made to the diameter which gives the economic velocity of about 50 metres per minute. This figure has been found to hold true in several cases where rising mains had to take dry-weather flow. It should be stressed that the economic velocity remains the same regardless of the length of the rising main. In the case of pumping stations for storm-water where the maximum rate of flow through the rising main does not occur except at rare intervals, the economic velocity is higher and may be increased to any convenient figure provided it does not exceed the maximum of about 146 metres per minute. The economic sizes of large rising mains should be calculated by trying various diameters and working out for each the annual cost of electricity plus the annual repayment of loan (see Table 20) on the capital expenditure on pumping station and rising main. It will then be seen that, as in the example given in Table 19, there is a diameter of rising main which calls for the minimum annual cost.

4

Pneumatic Lifting of Sewage and Sludge

Two types of air-operated devices are used for lifting sewage and sludges: these are sewage ejectors and low-lift air lifts.

Sewage ejectors

Sewage ejectors have been very largely used at small sewage-pumping stations dealing with flows up to about 1 cubic metre per minute (or more in some instances) against moderate heads. They are not efficient in terms of power required in proportion to work done, particularly at high heads, but they have the advantages of reliability and of keeping the rising mains clean. They will pump crude sewage and any sewage sludge that will flow freely under gravity.

A sewage ejector consists of a cast-iron body into which the sewage flows from a manhole *via* an isolating sluice valve and reflux valve. On the delivery side is another reflux valve and sluice valve isolating it from the rising main. Inside the ejector is a float (or, in some makes, an open-bottomed bell and an inverted bell) which operates an air valve. On the ejector being filled with sewage the float rises to admit compressed air which blows the sewage up the rising main. The float then closes the air inlet and opens an air outlet to permit further inflow of sewage. As the air from the outlet is foul, it is discharged by pipe to the sewer.

An ejector installation usually consists of two ejectors controlled by an automatic air valve to work alternately, one filling while the other is emptying. Both ejectors are connected to the same rising main. The air is supplied by a small compressor, or compressors in duplicate. Air is delivered to an air vessel, for the discharge of the ejector can be momentarily more rapid than the output of the compressor. The compressor is stopped and started by the limiting air pressures in the air vessel.

Most ejectors receive their inflow by gravity, but the Tuke and Bell 'Lift and Force' ejector can draw from a suction well at a lower level.

TABLE 21. *Compression of air*

Head (metres)	Pressure* (kilogrammes per square centimetre)	Cubic metres free air per cubic metre of compressed air	Kilowatts per cubic metre free air per minute (at 100% efficiency)
0.9	0.09	1.09	0.141
1.2	0.12	1.12	0.185
1.5	0.15	1.15	0.228
1.8	0.18	1.18	0.270
2.1	0.21	1.21	0.312
2.4	0.24	1.24	0.352
2.7	0.27	1.27	0.392
3.0	0.30	1.30	0.430
3.3	0.33	1.32	0.466
3.6	0.36	1.35	0.503
4.2	0.42	1.41	0.573
4.8	0.48	1.47	0.643
5.4	0.54	1.53	0.706
6.0	0.60	1.59	0.771
7.5	0.75	1.74	0.919
9.0	0.90	1.89	1.05
10.5	1.05	2.03	1.18
12.0	1.20	2.18	1.30
13.5	1.35	2.33	1.41
15.0	1.50	2.48	1.50
18.0	1.80	2.77	1.69
21.0	2.10	3.07	1.86
24.0	2.40	3.36	2.00
27.0	2.70	3.65	2.14
30.0	3.00	3.96	2.28

* These figures, multiplied by 98 066.5, give newtons per square metre or pascals.

Sewage ejectors are sold in terms of the capacity of the cast-iron vessel and this is often considered to be the quantity that can be delivered per minute. In fact, the delivery depends on the rate at which sewage can gravitate into the ejector, which will vary with the head on the inlet side, and the rate at which the sewage is blown out, dependent on the air pressure.

As in the case of sewage-pumping stations, the rising main should not be less than 100 millimetres in diameter and the air pressure should be sufficient to produce a velocity of not less than 46 metres per minute in this rising main. Rule-of-thumb practice is to have an air pressure 40% greater than the manometric head. The quantity of air required can be found by the formula

$$C = \frac{Q(H+10\cdot3323)}{10\cdot3323},\qquad(29)$$

where C = cubic metres free air per minute,
H = total manometric head in metres,
Q = discharge of sewage in cubic metres per minute.

Low-lift air lifts

Air lifts are best known for their use in pumping from wells for water supply purposes or for testing new wells. When lifting against high heads they are not mechanically efficient but they are very reliable. In sewage-works practice, air lifts can lift against very low heads, in which case it is found that the deep-well design formulae do not apply and a new empiric rule has to be used.

The sewage-works purposes for which low-lift air lifts are suitable are lifting detritus from detritus channels, recirculating activated

TABLE 22. *Optimum efficiency of low-lift air lifts (at optimum velocity and excluding efficiency of compressors)*

Submergence ratio	% efficiency	Submergence ratio	% efficiency
1·5	21·15	4	27
2	22·8	4·5	27·8
2·5	24	5	28·6
3	25·2	5·5	29·2
3·5	26	6	29·9

TABLE 23. *Performance of low-lift air lifts: approximate cubic metres of air per cubic metre of water lifted*

Submergence ratio	Velocity of water in metres per second										
	0·30	0·45	0·60	0·75	0·90	1·05	1·20	1·35	1·50	1·65
1·5	4·54	3·98	3·58	3·30	3·16	3·16	3·32	3·60	4·03	4·60
2·0	3·18	2·78	2·50	2·30	2·22	2·22	2·32	2·50	2·82	3·20
2·5	2·40	2·10	1·89	1·74	1·67	1·67	1·75	1·90	2·13	2·42
3·0	1·91	1·68	1·50	1·39	1·33	1·33	1·39	1·51	1·69	1·93
3·5	1·58	1·38	1·24	1·15	1·09	1·09	1·15	1·25	1·39	1·59
4·0	1·33	1·17	1·05	0·96	0·93	0·93	0·97	1·06	1·18	1·34
4·5	1·15	1·01	0·90	0·83	0·80	0·80	0·84	0·92	1·02	1·16
5·0	1·01	0·88	0·79	0·73	0·70	0·70	0·74	0·80	0·90	1·02
5·5	0·90	0·78	0·70	0·65	0·62	0·62	0·65	0·71	0·79	0·90
6·0	0·80	0·70	0·63	0·58	0·56	0·56	0·58	0·63	0·71	0·82

sludge, lifting effluent, etc. They are equally effective for pumping clear water and heavy suspensions.

An air lift consists of a vertical pipe deeply submerged in the liquid to be raised. Air is admitted at the bottom end, and the bubbles which rise up the tube lighten the column of water so that it is lifted above the level of water in the suction well. There are several arrangements. The air pipe can pass down inside the vertical tube to discharge air just above the inlet at the bottom. Another method is to take the air down outside and introduce it through holes in a collar round the bottom of the pipe. A third is to have the uptake tube inside a larger tube containing compressed air: this has some practical uses but is not common.

The efficiency of low-lift air lifts depends mainly on the submergence ratio, i.e. the depth to which the tube is submerged divided by the dead lift. It also depends on the velocity in the uptake tube, there being an optimum velocity which, in the experiments on which Tables 22 and 23 were based, was almost exactly 1 metre per second. Thus the delivery of a low-lift air lift can be taken, for design purposes, as being 1 cubic metre per second per square metre of waterway (excluding any space occupied by an internal air pipe).

The bottom of the air lift should be shaped to admit water freely; there should be free discharge at the top, and also a deflector plate to prevent the air lift from throwing spray to a height. If the air lift

TABLE 24. *Approximate efficiencies of low-pressure air blowers*

Capacity of compressor (cubic metres of free air per minute)	Overall efficiency (%)
28·4	53*
56·7	58*
85	62*
113	64*
140	66*
170	71†
198	72†
226	73†
284 upwards	74†

* Roots-type blowers.
† Turbo blowers.

discharges below the water surface, the depth below the surface must be deducted from the submergence in the calculations. Table 22 gives the approximate efficiencies of low-lift air lifts and Table 23 gives the quantities of air required for operation.

The degree of pressure in the air supply requires to be slightly in excess of that necessary to blow the air down to the submergence of the air pipe. Thus the necessary power can be calculated from Table 21. See Table 24 for efficiencies of air blowers.

5

Construction of Sewers

Nearly all pipes found in nature are of circular cross-section which is the natural shape caused by internal pressure and is also best able to resist external pressure. Similarly, in engineering construction, the vast majority of pipes are manufactured to circular cross-section, and brick and concrete culverts are more frequently circular now than formerly. Once, large culverts could be built more economically with vertical sides and flat tops, but the introduction of special steel shutters that could be moved forward with the work and used many times has made curved work no more expensive than straight work.*

In addition to special shapes intended to reduce costs, there were cross-sections used because it was fallaciously believed that they favoured self-cleansing conditions when gradients or flows were inadequate. The best known of these was the 'old' or 'standard' egg-shaped sewer† which had a depth of three times the crown radius and an invert radius of one-half the crown radius. Many ancient sewers of these shapes still exist, but cross-sections of other than circular form are now obsolescent except when lack of headroom or some similar special circumstance enforces variation from the circle.

Nominal metric dimensions

In spite of the change from imperial to metric measurement there are

* In the recent construction of some very large-diameter reinforced-concrete culverts the contractor requested permission to modify the shape, from true circle to flat invert and 45° slopes from the invert tangential to the circular barrel, so as to economize in cost of placing concrete in the invert. After laying the first length he reverted to the circular cross-section because he found it equally easy to construct.

† The 'new' egg-shaped sewer, used in the nineteenth century, went out of fashion earlier than the 'old' type.

TABLE 25. *Vitrified clay pipes with flexible joints*

Manufacturer	Available sizes (millimetres diameter)	Type	Trade name
Church Gresley Fire Brick and Fire Clay Co. Ltd	100, 150, 225	Polyester	C. G. Swadflex
John Crankshaw Co. Ltd	375, 450	Polyester	Hepseal
Donington Sanitary Pipes and Fire Brick Co. Ltd	100, 150, 225	Polyester	Donflex
Dorset Clay Products Ltd	100, 150, 225, 300	Polyester	D. C. Polyester Joint
Doulton Vitrified Pipes Ltd	100, 150, 175, 225, 300	Polyurethane	Draw Flex
Ellistown Pipes Ltd	100, 150, 175, 225, 300	Polyester	Ellflex
Eltringham Pipe Co. Ltd	100, 150, 225, 300	Polyester	Polyester
Hawfields Brick and Pipe Works Ltd	100, 150, 225	Polyester	Hawseal
Hepworth Iron Co. Ltd	100, 150, 175, 225, 300, 375	Polyester	Hepseal
Hepworth Iron Co. Ltd	100, 150	Hepsleve	Hepsleve
Kinson Pottery Ltd	100, 150, 225	Polyester	Kinflex
John Knowles & Co. (Wooden Box) Ltd	100, 150, 225	Polyester	Vitriflex

CONSTRUCTION OF SEWERS

Manufacturer	Sizes	Material	Trade Name
W. T. Knowles & Sons Ltd	100, 150, 225, 300	Polyester	Knolflex
Lochside Coal & Fire Clay Co. Ltd	100, 150, 175, 200, 225, 300, 375, 450	Polyester	Hepseal
Naylor Bros (Clayware) Ltd	100, 150, 175, 200, 225, 300, 375	Polyester	Naylor Polyester Joint
James Oakes & Co. (Riddings) Ltd	100, 150, 225, 300, 375, 450	Polyurethane	Fast-test
James Oakes & Co. (Riddings) Ltd	100, 150, 225, 300, 375, 400	Hot pour	Oanco
Sneyd Pipeworks Ltd	100, 150, 300	Polyester	Hepseal
Stella Tileries Ltd	100	Polyester	Stella Tileries Hepseal
Sutton & Co. (Overseal) Ltd	100, 150, 175, 225, 300	Polyester	Ellflex
Western Pipes Ltd	100, 150, 225	Plasticized P.V.C.	Draw Flex
Woodville Sanitary Pipe Co. Ltd	100, 150, 175, 225, 300	Polyester	Ellflex
Thos Wragg & Sons Ltd	100, 150, 225, 300	Polyester	Easilay

many existing sewers constructed to imperial dimensions which, when they have to be repaired, may call for the use of pipes made to imperial dimensions. On the other hand, pipes are manufactured in some materials to nominal metric diameters which are interchangeable with pipes made to imperial dimensions. For example, asbestos-cement sewer pipes are sold in both imperial and nominal metric sizes, the nominal metric size being based on the convention that 1 inch equals 25 millimetres, not the actual 25·4.

Other materials will be manufactured to both imperial and metric sizes for many years to come. This will probably apply to cast-iron pipes. It should be observed that, while the smallest diameter of sewer, 100 nominal millimetres, is considered the same as 4 inches or 101·6 actual millimetres diameter, it would not be possible to joint together two large cast-iron pipes of, say, 48 inches and 1200 millimetres diameter.

Pipe sewers

For many years the materials most commonly used for the construction of small-diameter drains and sewers were salt-glazed ware pipes and vitreous-enamelled fireclay pipes. These are giving place to vitrified clay pipes which are not glazed but are available in longer lengths and are manufactured of more dense material to closer tolerances. These new pipes can be jointed in cement mortar in the same manner as the earlier stoneware pipes, but they are also available with special joints as listed in Table 25.

While glazed stoneware pipes have been made in large diameters, it will be observed from Table 25 that vitrified clay pipes are mainly limited to 100, 150, 225 and 300 millimetres. Above these sizes, concrete pipes with socketed or ogee joints are often preferred, while asbestos-cement sewer pipes are available in sizes from 100 to 900 nominal millimetres, and pre-stressed concrete pressure pipes have been made in various sizes from 300 millimetres diameter upwards.

According to size and circumstance, vitrified clay and concrete pipes with cement joints are laid with or without concrete protection. For many years, under Ministry recommendation, practice has been for pipes with more than 4·25 metres earth cover to be bedded on 150 millimetres of concrete and haunched with 150 millimetres of concrete to at least the horizontal diameter, the concrete being splayed tangentially to the pipe above that level, and for all pipes having more than 6 metres of cover to be surrounded with at least 150 millimetres thickness of concrete. All pipes of 450 millimetres diameter and over should be benched and haunched as described above, and all pipes with less than 1·2 metres cover when under roads or 1 metre elsewhere should be surrounded with at least 150 millimetres of concrete.

Apart from the foregoing, pipes of less than 450 millimetres diameter can be laid carefully bedded on undisturbed earth without concrete protection. It is, however, usual to put some concrete under pipes if the nature of the ground justifies this precaution or if, as is becoming more frequent than formerly, the standard of workmanship that will ensure sound and even bedding on earth cannot be obtained. Laying cement-jointed pipes on special granular fill is not to be recommended: all granular material, no matter what the quality, will subside to some extent unless some cement is added.

Some less definite recommendations have been made for the protection of sewer pipes with flexible joints and, as these recommendations are already considered obsolete and no substitute is as yet available, comment thereon will be reserved.

The practice of some engineers has been to use cast-iron pipes in lieu of clay or concrete pipes where conditions would require concrete surround for the latter materials. There is little difference in cost between a small-diameter vitrified clay pipe surrounded with concrete and an unprotected Class 'B' spun-iron pipe, but there are differences in advantages. The cast-iron pipe sewer will not develop leaks permitting appreciable infiltration, as will a sewer of cement-jointed clay or concrete pipes, and for this reason there are several cases when it is to be preferred. On the other hand, should the sewer in question be one to which several further lateral connexions may have to be made, cast-iron is an expensive material, for the correct way of making a new lateral connexion to cast-iron is to cut out a section of pipe, insert a junction pipe with two collar joints, and make good.

Pipelines to take internal pressure

Sewage rising mains, inverted siphons and all sewers and drains liable to be under internal pressure sufficient to damage a vitrified clay, concrete or asbestos-cement sewer pipe, may be constructed of spun-iron, vertically-cast iron, malleable cast-iron pipes, steel pipes and asbestos-cement pressure pipes.

The cast-iron pipes used for sewerage purposes are not of the same type as cast-iron drain pipes (B.S. 437) but are similar to those used in water supply (B.S. 78 and B.S. 1211) and, except when the internal pressure justifies a heavier grade, Class 'B' pipes are used. The cast-iron pipe industry manufactures grey-iron pipes according to the current international metric standard of the International Organization for Standardization, Recommendation R.13, as well as the British Standard imperial sizes. The industry has also changed to fittings of the all-socket type as used in America and Europe, and these will eventually replace spigot-and-socket fittings.

At one time it was usual for cast-iron pipes laid in trench to be jointed with molten lead caulked by hand. The practice now is to use bolted-gland joints with considerable saving in lead. These are satisfactory in trench but cannot be used in any place where pipes are unsupported, for the joints easily blow out, whereas the ultimate strength of caulked lead joints is understood to be 14 kilograms per square centimetre of lead in contact with the outer surface of the spigot, which gives reasonable resistance to blowing out in the case of small-diameter pipes.

When cast-iron pipes are carried above ground level, they should be supported behind every joint unless they are flanged pipes, in which case it is understood that the following formula gives approximately the safe span:

$$L = \tfrac{1}{2}D^{0.5}, \tag{30}$$

where L = maximum span in metres
D = internal diameter in millimetres.

(Most engineers will probably prefer to reduce the span so calculated.)

Large-diameter sewers

Sewers of large diameter, say 750 millimetres or more, may be constructed of concrete pipes surrounded with concrete but, in the larger diameters, concrete pipes become an expensive form of shuttering and, unless only a short length of sewer is envisaged, construction in mass concrete is more economical. When mass concrete is used the centre of the invert may, with advantage, be lined with very dense vitrified brickwork as a protection against scour. As this scour is mostly caused by large detritus which slides down the sewer forming grooves, it is not necessary for more than about one-sixth of the circumference to be protected in this manner.

In these larger diameters it may also be necessary to design the thickness of the walls of the sewers according to the calculated external earth or water pressure. These external pressures are always speculative, for they depend on many factors, including the method of excavation, which may not always be that which has been allowed for in design. If a sewer is laid in narrow trench it is probable but not certain that the pressure of earth upon it will be considerably less than the weight of the superincumbent earth because of friction between the fill and the sides of the trench. But the designer is not to know whether or not the contractor will excavate a narrow timbered trench or a wide ditch with sloping sides: in the latter case the vertical pressure may be virtually equal to the weight of superincumbent earth.

It is frequently overlooked that external water pressure, which in porous ground can be from the highest natural water table (or subsoil water level) and in heavy clay or other impermeable subsoil can well extend to ground level, may exceed the earth pressure. It is therefore advisable to calculate the maximum vertical earth pressure bearing on the top of the sewer and the maximum water pressure crushing it from all sides, and to make sure that the sewer will withstand each or both of these. (It should be mentioned that no structure designed by the author, on the basis of water pressure only, has been known to fail.)

For rigid sewers, i.e. those not having flexible joints, on ordinary bedding and not occupying less than two-thirds of the trench width, the probable superincumbent earth load can be calculated by the formula

$$W = 2000 H^{2/3} B^{4/3}, \qquad (31)$$

where W = vertical pressure in kilograms per metre run of sewer,
H = depth of fill over top of sewer in metres,
B = width of trench at top of sewer in metres. (In no case can B exceed H.)

Some guidance to the required thickness of brick-barrel sewers to withstand water pressure is given in Table 26.

This table is based on the assumption that the sewer is subjected to an external pressure equal to a head of water from ground level to the centre of the sewer, and also that in no case is the thickness of the barrel less than one-twelfth the internal diameter.

If a pipe is supported on piers or on piles and beams above

TABLE 26. *Maximum depth in metres of brick-barrel sewers (ground level to centre of barrel)*

Internal diameter (nominal millimetres)	Thickness of barrel (nominal millimetres)			
	113	225	340	450
750	15	—	—	—
825	12	—	—	—
900	10	32	—	—
975	8	30	—	—
1050	6	27	—	—
1125	4	24	—	—
1200	—	22	—	—
1275	—	20	—	—
1350	—	18	32	—
1425	—	16	30	—
1500	—	15	29	—
1650	—	12	25	—
1800	—	10	22	32
1950	—	8	19	30
2100	—	6	17	27
2250	—	4	15	24
2400	—	—	13	22
2550	—	—	11	20
2700	—	—	10	18
2850	—	—	8	16

natural ground level but is buried in fill, the subsiding fill slides past the supported sewer, which then has to carry the weight of a wedge of earth, this weight can be very great indeed and, unless allowed for, can cause both beam and pipe to fail. There is no satisfactory theory covering this case and laboratory experiments carried out for the author gave widely varying results when conditions were apparently the same. On the basis of these experiments, it was decided that concrete beams carrying pipes should be designed to take the weight of wedges of earth that began at the bottom with a width equal to the widest part of the pipe or beam and sloped outwards at each side at an angle of 15° to the vertical up to formation level. This rule was used with the added proviso that the stress in the reinforcement should not exceed 2100 kilograms per square centimetre or 200 newtons per square millimetre under the load of a wedge of earth having sides sloping at 30° to the vertical, the maximum experimental angle. So far no failure has been reported of structures designed to this rule. When pipes are laid on beams in this manner, they must be grouted up from the beam throughout their lengths, and not supported at intervals only.

Large sewers in tunnel

When large sewers have to be constructed at such depth or in such circumstances that it is advisable to work in tunnel, it often proves economical to use segments of cast-iron or reinforced concrete which can be erected as the work proceeds, serving the purposes of preventing the earth from falling in during construction and providing the permanent structure of the sewer. Pre-cast concrete segments bolted together with bitumen packing between the joints and lined with mass concrete are available in several designs, some of the best known being those made by Kinnear Moodie & Co.

Both concrete and cast-iron types of construction are suitable for use with the aid of a shield and under compressed air, but cast-iron segments are to be preferred where the external ground-water pressure exceeds 9 metres and it would be difficult to make watertight joints with concrete segments.

Cast-iron segment sewers consist of a number of rectangular segments bolted together to form rings which, in turn, are bolted together by flanges on the inside. The flanges that run longitudinally are radial except for those of a small 'key' segment which is placed in the crown to complete the ring and, of course, the adjacent flanges of the 'taper' segments on each side of the key segment. The key segments are, of necessity, tapered, being 254 millimetres wide at the outside of the ring and 260 millimetres at the inside. The length of each completed ring is 508 millimetres in which there are three bolts per longitudinal joint. Each cast-iron segment has a grout hole of about 32 millimetres diameter.

There are no standard dimensions for cast-iron sewer segments, neither are there any catalogue sizes, but Table 27, prepared by interpolation between the dimensions adopted in a number of London County Council contracts, will give guidance for design purposes.

TABLE 27. Dimensions of cast-iron segmental sewers (millimetres)

Internal diameter of concrete lining	1525	1600	1675	1830	1980	2135	2285	2440	2590	2745	2895	3050
Internal diameter of iron	1651	1727	1803	1956	2108	2261	2413	2565	2718	2870	3023	3175
External diameter of iron	1855	1931	2007	2160	2324	2477	2629	2793	2946	3112	3265	3417
Thickness of iron	19	19	19	19	19	19	19	19	19	19	19	19
Depth of flanges	83	83	83	83	89	89	89	95	95	102	102	102
Thickness of flanges at base	25	25	25	29	29	29	29	29	29	29	29	29
Thickness of flanges at edge	22	22	22	25	25	25	25	25	25	25	25	25
Diameter of bolt circle	1734	1810	1886	2039	2197	2350	2450	2660	2813	2972	3125	3277
Number of bolts	19	23	23	23	23	27	29	29	34	36	39	39
Size of bolts, diameter	19	19	19	22	22	22	22	22	25	25	25	25
Size of bolts, length	100	100	100	100	108	108	108	108	108	108	108	108
Number of ordinary segments (excluding one key and two taper segments) per ring	3	4	4	4	4	4	4	4	5	5	5	5

CONSTRUCTION OF SEWERS 85

To form the rings, the longitudinal joints, which have machined faces, are bolted together with a suitable jointing material such as red lead and lead wire and, if this does not produce a watertight joint, they are caulked with lead wool in a 7-millimetre groove, 18 millimetres deep, on the inside, which may be finally pointed with rust cement, a mixture of cast-iron borings and sal-ammoniac in the proportion of 400 to 1. If the work is in a straight line, the circumferential joints are bolted together, adjustment being made as necessary to correct any errors of line and gradient. Also slight curvature can be effected without the use of special segments. If the work is in the dry, the circumferential joints are caulked with creosoted deal packing; otherwise lead wire and red lead, followed by rust cement, are used. In erection it is usual to stagger the key segments to break joint, but this is not possible when special segments are used on curves: then the key segments all come at the centre at the top of the arch.

Connexions of lateral pipes to cast-iron segmental sewers are made by special castings shaped to bolt to the curved faces of the segments which are cut or burned out as required.

The lining of a cast-iron segmental sewer usually consists of mass concrete shuttered on the inside and, preferably, with a hard vitrified-brick invert.

Testing drains, sewers and rising mains

Drains and sewers which are not intended to function under pressure are tested by the air test, the water test, or both. The air test is the most convenient for testing pipes that are exposed before refilling of earth, because it avoids the necessity of providing and disposing of large quantities of water. It can be useless after refilling for, if the sewer is submerged in subsoil water by as little as 100 millimetres, no air will be lost during the test no matter how leaky the sewer may be. The air test also has the advantage in that it is applicable to long lengths of steeply-falling sewer which, if water-tested, would have to be tested in successive short lengths to avoid developing excessive pressures at the lower end. (These should never exceed 6 metres head of water.)

The water test is the most usual for testing sewers after the trench has been refilled. However, most engineers consider that, if a sewer has been properly laid, has satisfactorily passed an air test and the refilling of earth has been completed properly, a second test is not required.

The air test is applied by closing all inlets and outlets to the sewer with expanding plugs or bags, and pumping in air to a pressure of 100 millimetres of water-gauge as recorded on a U-tube situated at the opposite end of the pipeline from that at which the air is introduced. The pressure must then not fall more than a specified amount during a specified period of time, after which the pressure is released at the pump end of the sewer, when the drop of the pressure on the U-tube should then indicate that the test has not been falsified by a secret plug.

In the water test all outlets are similarly plugged except at the top end, where a temporary bend and upstand pipe of the same diameter and material of the sewer are fixed in such a manner that a minimum head of water of 1·2 metres can be applied above the crown of the pipe at that end. The pipe is then left to soak for a period of at least 1 hour, or longer if the contractor so wishes, after which the water level is topped up and the fall by leakage (if any) recorded. It is important for the upstand pipe to be of the same diameter as the sewer, as otherwise air will be trapped in the sewer, falsifying the test: also, a different diameter makes calculation of the permissible amount of drop more difficult.

The author recommends the durations of test and permissible falls of water level given in Table 28, for salt-glazed pipe, vitreous enamelled pipe, asbestos pipe or cast-iron pipe sewers jointed in cement mortar. For slightly porous materials, such as vitrified clay pipe or concrete pipe sewers, the severity of the air test only may be reduced to one-half.

Lead-jointed cast-iron drains or sewers that will not function under pressure may be tested by air or water test, but no leakage whatsoever may be permitted. There are other tests such as the smoke test used for testing old drains: these are not now considered suitable for new sewers.

Rising mains, inverted siphons and all other drains or sewers that will function under pressure are given hydraulic tests. All ends and openings are sealed off and strutted to resist pressure. All bends and junctions are concreted to the side of the trench and the middle of each pipe is covered with earth to prevent movement, the joints being exposed for inspection. Water is then pumped into the pipe until a pressure of not less than one and a half times the working

TABLE 28. *Air and water tests for sewers and drains*

Internal diameter of pipe (millimetres)	Air test		Duration of water test for 50 millimetres max. drop (minutes)							
	Duration (minutes)	Maximum water drop (millimetres)	$2\frac{1}{2}$	5	$7\frac{1}{2}$	10	15	20	25	30
			Minimum length of pipe under test (metres)							
75	$1\frac{1}{2}$	25	60	30	20	15	10	$7\frac{1}{2}$	6	5
100	2	25	80	40	27	20	13	10	8	$6\frac{1}{2}$
150	3	25	120	60	40	30	20	15	12	10
225	$4\frac{1}{2}$	25	—	90	60	45	30	$22\frac{1}{2}$	18	15
300	5	20	—	120	80	60	40	30	24	20
375	6	20	—	—	100	75	50	$37\frac{1}{2}$	30	25
450	$5\frac{1}{2}$	15	—	—	120	90	60	45	36	30
525	$4\frac{1}{2}$	10	—	—	—	105	70	$52\frac{1}{2}$	42	35
600	5	10	—	—	—	120	80	60	48	40
675	$5\frac{1}{2}$	10	—	—	—	—	90	$67\frac{1}{2}$	54	45
750	6	10	—	—	—	—	100	75	60	50

pressure has been attained, and this pressure should be maintained without any visible fall for a period of at least one half-hour without further pumping.

6
Sewer Appurtenances

Manholes

Manholes give access to small sewers for rodding or flushing* to clear stoppages and means of entry to those large sewers through which men can walk. They are designed to economically permit easy and safe access.

A manhole consists of a chamber in which men can work and, if the manhole is deep enough, a shaft leading down to the chamber. In some cases where manholes are placed in the carriageways of roads for heavy traffic there may be galleries leading laterally from shafts under the footways. Each manhole has a cover suitable for the load to which it may be subjected, a ladder or step-irons leading down, properly formed inverts to take the flows, and benchings to give standing positions for the men.

Very small manholes used on private drains and of insufficient depth for men to enter are known as inspection chambers. These are different in dimensions and type of construction from public-sewer manholes. They may have as little as 0·45 to 0·6 metre depth from ground level to invert, in which case they need to be designed to permit rodding from ground level. When on small drains and with no side connexions, they may be as small as 0·7 by 0·45 metre on plan but, when the depth exceeds 0·6 metre, manholes measuring not less than 1·25 metres long by 0·9 metre wide are preferred for good quality work. Where there are many side connexions, the size of the chamber may have to be increased to accommodate the stoneware channel bends in the benchings. Inspection chambers or manholes that are not more than 1 metre deep do not require ladders or step-irons.

Manhole chambers that are in the region of 1·5 metres deep are

* Sewer men are still known as flushers from the days when the cleansing of large, ill-designed sewers was a major part of their work.

the most difficult to work in, for they are too deep to make easy rodding from ground level and too shallow for a man to enter and rod from benching level. These need to have extra-large manhole covers centrally located between the two ends of the chamber.

The vast majority of manholes are medium-depth manholes averaging in the region of 2 metres deep. In the contract drawings for most sewerage schemes it is necessary to have standard designs for shallow, medium-depth and deep manholes. The shallow manholes are generally as described above, the medium-depth manholes have chambers measuring 1·37 metres to 0·9 metre on plan (or thereabouts to the nearest brick dimension) and have no shaft. The deep manholes have chambers of similar size but of heavier construction to withstand greater external pressure, and shafts leading to ground level.

The invert of a manhole is usually constructed of mass concrete and lined with material of the same kind as used for the sewer. In its thinnest part it does not need to be thicker than 150 millimetres or the thickness of the sewer, whichever be the greater. The invert should be rounded to the radius of the sewer up to half diameter, brought up vertically to soffit level, then turned over sharply to form the benching which should slope at about 1 in 6 towards the channel. A steeper slope is dangerous: too slack a slope will not drain well enough.

Wherever practicable, the chamber of a manhole should give a headroom of 2 metres from the surface of the benching to the underside of the roof. The consensus of opinion is that rectangular access shafts should measure 0·7 metre by 0·8 metre, the larger dimension being from the wall to which the ladder is fixed. A smaller dimension than 0·8 metre can be cramping; a larger dimension does not permit a man to rest his back against the wall should he need to do so. When shafts are circular, a diameter of 0·7 metre is frequently used because this is standard for certain makes of pre-cast manholes, but 0·75 metre is to be preferred.

Manhole ironwork

Permanent ladders or step-irons are necessary in all manholes except those which are so shallow as to make them unnecessary. Step-irons are often used because they are inexpensive, but they have the disadvantages of being liable to break off (if made of common cast-iron), of wearing and developing a sharp edge that will cut the hand, and of corroding away until they are dangerous. In the last condition they become expensive to cut out and replace.

Step-irons also have the disadvantage that they are almost invariably wrongly installed and it can be assumed that, if their insertion is left to the bricklayer, they will be arranged in such a manner as to be inconvenient or dangerous.

Step-irons should be built-in at regular intervals of 0·3 metre (or if preferred by the designer 0·225 metre) vertically, staggered regularly to left and right. They should never be staggered more than 0·2 metre apart, centre to centre, as this makes them hard to find with the feet and causes the sewer man to sway from side to side. They should never be out of line, irregularly spaced or missing. All step-irons should be of unbreakable material such as galvanized malleable cast-iron. They should never be used in manholes that are frequently entered, and each should be tested by being given a sharp blow with a 2-kilogramme hammer. Generally they should not be used for vertical flights of more than 2·75 metres.

Galvanized mild-steel ladders are used for most manholes. They are usually 0·3 metre, but may be 0·375 metre wide: excessive width can be inconvenient. The rungs are usually 0·3 metre apart vertically, but 0·225 metre is to be preferred for ladders that are to be used by men wearing waders or sewer boots. To give a good foothold, the rungs should be about 0·225 metre from the face of the wall to which the ladder is fixed.

Galvanized ladders should be fixed to walls with gunmetal bolts or other means which will not involve the labour of cutting out rusty stubs of iron when the ladders have to be replaced. Galvanized steel ladders will last for a long time in conditions where they are occasionally flooded by sewage, for the greasy sewage has a protective effect: it is above top sewage level that rusting due to condensation is most likely to take place. Aluminium ladders have the disadvantages of being of lesser strength, too springy a nature and involving higher cost than galvanized steel of the same cross-section: generally they are not to be recommended.

When manholes are very deep, it is desirable to have intermediate chambers at distances not exceeding 6 metres vertically. In these chambers there should be platforms measuring not less than 0·9 metre by 1·4 metres on plan. They should be fitted with hand-rails and hinged gratings so arranged that building materials may be lowered or hoisted vertically from the surface or casualties lifted with the aid of safety harness.

Gratings are best constructed of galvanized steel consisting of bars of not less than 50 by 9·5 millimetres cross-section, tenonned together: no butt-end electric or oxy-acetylene welding should be permitted.

Manhole covers for use in roads are usually made of common cast-iron. So that they shall not interfere with rescue operations using a safety harness, they should have circular openings of not less than 0·56 metre diameter: those on large sewers may have greater diameters to facilitate the lowering of special skips, but in all cases the plug should not weigh more than 140 kilogrammes. The cover should be capable of taking a test load of 35 600 kilogrammes applied through a 300-millimetre diameter hardwood block. The author does not favour triangular covers for use on main sewers because none of the well-known designs has a sufficiently large opening for easy rescue.

Drop manholes

In nearly every contract there is a circumstance in which, for reasons of economy or otherwise, sewage has to be dropped from a high to a low level. The most frequent arrangement is a 'backdrop' which consists of a vertical pipe surrounded by concrete outside the manhole and brought in with a bend at an angle to the direction of flow. Often a cast-iron bend with special socket is set in concrete at the bottom: the vertical pipe is carried in the same material as the sewer to ground level, where a lamphole cover gives access for rodding. The sewer itself is carried across the backdrop to the shaft or chamber of the manhole where a tide flap gives access for rodding. The foregoing type of construction has taken the place of a less satisfactory, more expensive 45° 'ramp' which was used early in the century.

Vertical drops consisting of cast-iron soil pipes (B.S. 416) are used in manholes on private drains. The incoming high-level drain connects to a cast-iron bend inside the manhole. Thence a cast-iron soil pipe continues down to the invert where a bend is fixed, turning in the direction of the flow. The upper bend has a bolted access permitting rodding in both directions. The whole is jointed with caulked lead joints and fixed to the manhole wall.

Drops on large sewers can consist of backdrops as already described but with water cushions at the bottom. These are chambers which will hold a sufficient body of water to dissipate the energy of the fall. In many instances multi-stage backdrops have been built, but a single-stage backdrop is adequate for dealing with very deep drops provided that the vertical shaft is of such size that the flow either falls clear through the air or clings to the side, not filling the pipe.

Head can be dissipated inside large-diameter sewers by cascades or tumbling bays in the form of flights of steps. Wherever practicable these steps should be of such proportions that they can serve as stairways for sewer men, each step being not more than 0·225 metre high. The necessary width of the cascade can be calculated by the formula

$$B = Q/100\ H^{3/2}, \qquad (32)$$

where B = width of cascade in metres,
  Q = cubic metres per minute,
  H = approximate depth of flow (one-half the fall of one step) in metres.

Storm overflows

Storm overflows were often constructed to discharge untreated stormwater from combined sewers to natural watercourses, and various rules-of-thumb and theories of design were formulated. This practice is now prohibited by Sections 30 and 31 of the Public Health Act, 1936 and storm water must be given at least partial treatment in storm tanks before discharge. For this reason the subject will be discussed later under the heading of storm-water separation.

Construction of manholes, chambers and soakaways

Manholes in towns are most frequently constructed of brickwork with either arched or reinforced concrete roofs and mass concrete inverts, benchings and foundations to walls. Generally the foundations should not project beyond the walls as this involves extra cost to no advantage. When shafts or chambers are corbelled, each corbel is not more than 30 millimetres per course of brickwork.

Bricks used for all sewerage purposes should be dense, vitrified bricks that will not absorb water by more than 4% of their dry

TABLE 29. *Thickness of manhole walls. Approximate maximum depth in metres for internal or external water pressure on manhole walls*

Length of internal face of longest wall (metres)	Thickness of wall (millimetres)							
	225	338	450	563	675	788	900
1	4·4	10·0	17·6	—	—	—	—
1¼	2·8	6·4	11·3	17·6	—	—	—
1½	2·0	4·5	7·8	12·2	17·6	—	—
1¾	1·4	3·3	5·8	9·0	13·0	17·6	—
2	—	2·5	4·4	6·9	10·0	13·5	17·6
2¼	—	2·0	3·5	5·4	7·9	10·6	14·0
2½	—	1·6	2·8	4·4	6·4	8·6	11·3
2¾	—	—	1·5	3·6	5·3	7·2	9·3

weight when broken and soaked for 24 hours. They should be laid frog upwards in rich cement mortar and flushed up or grouted so as to produce completely watertight work. Table 29 gives thicknesses of walls for manholes and chambers, constructed of brickwork in cement mortar or mass concrete of ordinary quality, to withstand the equivalent of water pressure from ground level either inside or outside the manhole.

Mass concrete may also be used for manholes and can have good appearance if the shuttering and workmanship are satisfactory. As much concrete as possible should be poured at one time, and any concrete that has set or partially set should be thoroughly cleaned off to expose the aggregate before more is deposited. Vibrating is a means of ensuring watertight work.

Apart from reinforcement of roof slabs there is very little advantage in using reinforced concrete for manholes, because manholes constructed in the manner described and to the proportions given in Table 29 are very strong and less liable to develop shrinkage cracks than reinforced work. Moreover, reinforcement is difficult to place in the restricted excavations made for manhole construction and consequently the work can be unduly expensive.

Pre-cast concrete manholes constructed of pre-cast chamber and shaft rings, which are to the proportions of standard concrete pipes, together with taper-pieces to connect between chamber and shaft and pre-cast cover slabs, are largely used for rural and estate-development sewerage. They are inexpensive and can be built very quickly. Although inverts and benchings can be purchased pre-cast to required form, it is more usual to construct these of mass concrete. It is advisable to set the manhole cover on one or two rings of brickwork to render adjustment easy should the road level be altered in any future reconstruction. (This precaution should also be observed when building mass or reinforced concrete manholes.)

Pre-cast concrete manhole segments can have either ogee or socket joints. It is usual to surround the chambers with concrete filled to the face of the excavation and brought up to the top of the chamber ring. Step-irons are provided, already built in by the manufacturer: they should never be used for hoisting purposes.

Manholes constructed of segmental cast-iron tubbing are manufactured for use where it would be difficult to make other forms of construction watertight. They are jointed in the manner already described for cast-iron segmental sewers.

Soakaways are usually circular on plan about 2·75 metres diameter. The walls are constructed to be of adequate strength yet to permit water to enter the chamber laterally. One method of construction is to have circular brickwork 225 millimetres thick in alternate rings of four courses in cement mortar and from one to four header courses laid dry with slightly open joints kept apart by pebbles. Another method is to construct the walls of brickwork in cement mortar with openings formed by the omission of headers at frequent intervals. The foundations of the walls are of mass concrete. The floors are best paved with bricks or slabs laid dry on the earth to render cleaning easy. Sometimes a layer of sand is placed on top of the floor. The roof may be formed of corbelling or a reinforced concrete slab with a shaft, terminating with a manhole cover, constructed to normal manhole standards. The surface water is brought into the soakaway preferably by a backdrop with a water cushion at the bottom to prevent undermining of the soakaway chamber, for the drains are usually at a very high level relative to the chamber bottom. The ends of the drains should be built through the walls as in backdrop manholes to permit rodding.

When excavating for a soakaway, work should proceed to the maximum depth intended or to the level of the subsoil water, whichever be the higher.

Arrangements for sewer flushing

The manhole at the top end of every lateral sewer should be provided with a flushing valve so that the manhole may be filled with water which may then be released to flush the sewer. A permanent mark should be provided in a visible position to show the highest level to which the water may be filled without causing backing-up or flooding of nearby drains.

Where a sewer has been laid at an unavoidably flat gradient or is too large for the expected flow, it may be desirable to construct an automatic flushing tank at the top end. Such a tank should have a capacity equal to about one-tenth of the cubic capacity of the length of sewer to be flushed. It should be connected to the sewer by an automatic flushing siphon, care being taken in the design of the manhole that the flow of air in and out of the flushing tank and adjacent manhole is free and will not interfere with the action of the siphon. If the flushing tank is fed by company's water, there must be a free fall from the incoming connexion from the water main into a small chamber that is isolated from the flushing tank by a trap which will not permit foul air to pass from the flushing tank.

Flushing has also been provided by controlled connexion to automatic tanks from natural watercourses or by pumping from wells. There are also means of flushing with sewage but these are now rare, except that a sewage pumping station at the head of a sewer almost invariably provides satisfactory flushing.

Ventilation

Sewers have to be ventilated for several reasons. First, it is necessary that positive or negative air pressure shall not interfere with the flow

TABLE 30. *Dangerous gases and vapours liable to be present in sewers*

Gas or vapour	Limiting toxicity concentration (% by weight)	Explosion limits (% by weight)
Acetone	0·1	2·6 to 13·0
Acetylene	—	2·5 to 80·0
Ammonia	0·03	15 to 26·0
Anaesthetics	0·1	—
Benzene	0·3	1·4 to 8·0
Carbon disulphide	0·1	1·0 to 50·0
Carbon dioxide	2·0	—
Carbon monoxide	0·01	12·5 to 74·0
Carbon tetrachloride	0·1	—
Chlorine	0·0004	—
Coal gas	0·01	5·0 to 31
Formaldehyde	0·003	—
Petroleum spirit	0·5	1·3 to 7·5
Perchlorethylene	0·1	—
Pyridine	0·005	1·8 to 12·5
Hydrogen cyanide	0·005	—
Hydrogen sulphide	0·002	4·3 to 46·0
Methane	—	5·0 to 15·0
Sulphur dioxide	0·003	—
Trichlorethylene	0·1	—

Note: These figures are taken from several sources. There are wide discrepancies and the safest value has been given in each case.

of sewage or cause the blowing or sucking-out of any traps on house connexions, gulleys, etc. Second, the gases which can accumulate in sewers are both poisonous and liable to cause damage. These include hydrogen sulphide which is highly poisonous, having caused many deaths of men employed in connexion with sewerage, and which destroys concrete. Petrol vapour is not only poisonous but explosive. Methane is explosive and causes oxygen depletion. Carbon dioxide causes oxygen depletion and has a dangerous toxic effect. Coal gas from leaky gas mains contains hydrogen and methane, which make it explosive, and carbon monoxide which makes it very toxic. All these gases and others that may be due to trade wastes are liable to collect in sewers unless ventilation is adequate.

Where there are no intercepting traps between drains and sewers the sewers are adequately ventilated because air can flow freely through the private drains and ventilating pipes at the heads of them. But in many parts of Great Britain it has been local practice to place intercepting traps between the drains and the sewers, and where this condition exists, it is necessary to make arrangements for the ventilation of the sewers.

The usual procedure is to provide a ventilating column at the head of every lateral sewer and to make sure that there is free flow of air at the bottom end of the main sewer where it discharges to a pumping station or sewage-treatment works. Ventilating columns or shafts consist of steel tubes with cast-iron bases set in concrete. The minimum internal diameter is 100 millimetres, but 150 millimetres is the usual minimum in practice and larger diameters up to 300 millimetres are catalogue sizes. Ventilating shafts can be purchased in heights ranging from 6 metres to 15 metres above ground level. The most usual catalogue size is 7·5 metres, but columns should always be tall enough to stand above the highest windows in the locality where they are to be used, and for residential areas, a height of 9 metres may be generally acceptable. Ventilating columns should be set truly vertical in 0·75 cubic metre of concrete filled to the face of excavation. The pipe which connects from the column to the sewer, because of its shallow depth at the top end, should be of cast-iron.

Ventilating manhole covers may be used where they will not cause nuisance, and are usual on surface-water manholes and soakaways.

House connexions

When new sewers are laid by a local authority for existing estates, the authority will have to make some provision for the lateral connexions of drains. This can be done by providing junction pipes in the necessary positions or by laying the lateral connexions as far as the boundaries of the properties to the requirements and cost of the property owners. It is usual to provide 45° junction pipes, as the sewer is laid using either Jenning's joinders, which have sealed caps that are easily cut off, or ordinary junction pipes with 100-millimetre branches temporarily blanked off. When the lateral connexion is made a bend is connected to the junction, and from this a straight line of pipe slopes up to the final manhole or inspection chamber on the private drain or private sewer. If the public sewer itself is constructed of cast-iron, the laterals will be of cast-iron also as far as the nearest private manhole.

When connexions are made to existing stoneware, clay or concrete-pipe sewers and no junction pipes have been left to take them, the sewer has to be cut by hammer and chisel, and a saddle, a device consisting of a flange, a short length of pipe and a socket is jointed to the sewer in cement mortar.

7

Coastal and Other Special Problems

Coastal towns do not have sewerage or sewage-disposal problems that are greater than those elsewhere. The contrary is true: they have the additional facility of the close proximity of the sea, to use or abuse, as a receptacle for sewage. Coastal towns may have separate, partially-separate or combined sewerage systems and surface water can be, and frequently is, discharged to soakaways. In some instances soil sewage is given full treatment, the effluent being passed to inland watercourses. But, because the sea is there, there may be partial treatment or no treatment at all.

The presence of the sea makes possible the discharge of surface water directly onto the foreshore, effecting some economy in sewerage costs where the separate system is used. It also suggests that, owing to the sea being a large body of water and therefore less liable to serious pollution than a small stream, sewage may be discharged into it with little or no treatment. Such discharge is made, and to a greater extent than is justified. A questionnaire issued in 1938 showed that, of the local authorities who replied, 54% discharged crude sewage without any treatment whatsoever, 26% discharged their sewage after screening or screening and by disintegration of the larger more objectionable solids so as to disguise but not remove them, 14% gave partial treatment and 6% full treatment. At the present time there has been some improvement but not to the extent desirable.

When untreated sewage is so discharged there is the danger of solids stranding on the beach: banks of sludge tend to accumulate at the end of the outfall; fat and grease float on the surface. There are places near well-known holiday resorts where one may stand on the shore and observe the sea opaque with sewage, or stand on a cliff and see the bathers in the greasy stream which is visible from a height but not from nearby.

While no case of illness has been proved to have resulted from contact with crude sewage in the sea, such contact must involve some danger to health, and sewage in the sea must always be considered undesirable at watering places. There is an additional fact which is not fully appreciated: when sewage enters fresh water it receives natural treatment and eventually becomes innocuous, but in the sea such biological and chemical changes are very much slower.

From the foregoing it would appear that coastal local authorities have to consider their means of sewage disposal according to local circumstances. If it can be proved that the tidal currents will carry the sewage far out to sea, then discharge of crude or disintegrated sewage may, perhaps, be acceptable. If there is a risk that, on some occasions, sewage may be brought to the shore in a well-diluted state, screening or screening and sedimentation may be sufficient precautions. But an important coastal resort which cannot risk damage to its reputation could well be advised to fully treat its sewage and discharge the effluent either to an inland watercourse or to the sea. Sewage treatment is never a heavy charge on the rates and, in such circumstances, could well pay for itself.

Sea outfalls

Sea outfalls for surface water are often outlets terminating in tide flaps and discharging onto the foreshore. Those for soil sewage are carried to below low water, spring tide, and are usually located at a headland because, where land juts out into the sea, the tidal currents are deflected away from the land. To gain some knowledge of what may happen it is necessary to make tidal experiments. The procedure is to place floats in the water at the proposed point of outfall at various times after high water over the entire tidal cycle and to follow them by boat, plotting their courses. This has to be continued until floats have been placed in the water at, say, 15-minute intervals for all times from high tide to 12 hours after high tide at both spring and neap tides, and the work takes several weeks. The floats can be very simple. Large oranges have been used, but the most satisfactory float is a piece of softwood, 75 millimetres square and 2 metres long, weighted with metal at one end so that it will float with about 0·5 metre freeboard, and the other end painted bright orange. Two floats should be used at a time.

Several surveying methods are used. One is to take angles with two theodolites from the shore to the boat (which must keep close to the floats) at exact intervals of time. But perhaps the simplest method is to read angles on a number of pre-erected shore stations from the boat with the aid of a nautical sextant. This method is based on the three-point problem which is as follows.

Let $P_1 P_2 P_3$ be three points observed on the land from the surveyor's boat and S stand for the surveyor's boat. The angles observed are $P_1 SP_2$ and $P_2 SP_3$, ϕ_1 and ϕ_2 respectively. Bisect $P_1 P_2$ and $P_3 P_4$ by perpendiculars A and B. These will pass through the centres of circles as shown on Figure 3. At P_1 or P_2 by means of protractor draw angle of 90° minus ϕ_1 and the line will pass through the centre of the circle which cuts $P_1 P_2 S$, because angle $P_1 O_1 P_2$ is equal to twice ϕ_1. Similarly find the centre of the circle which passes through $P_2 P_3 S$. The intersection of the two circles occurs at P_2 and S, thereby S is found. A number of different cases can occur, some of which are shown in Figure 3.

It will be observed how the above description applies to the cases in the upper part of Figure 3, where ϕ is less than 90°. In the left-hand part of the bottom of Figure 3, ϕ_1 is greater than 90° and therefore an angle of $\phi_1 - 90°$ has to be drawn on the opposite side of the cord $P_1 P_2$ from that on which $90° - \phi$ would have been drawn had ϕ been less than 90°.

The method of booking the results is as set out in Table 31. The results are plotted on an Ordnance Survey map with the aid of a station finder which is a protractor with three arms that can be set to the two angles ϕ_1 and ϕ_2. It will then be found that there is only one position of the point S at which the three arms will pass through the three shore stations except in the extremely rare circumstance when the two circles coincide. In the absence of a station finder, the angles can be drawn on tracing paper which can then be slid about until the point S is found.

When the survey has been plotted, showing the tracks of the floats started at various times after high water, it will become clear that there are times after high tide that should be suitable for discharge of sewage and others that are not likely to be suitable. If it is found that there is a time, or there are times, when the movements of the floats suggest that sewage can be discharged with little danger of its being washed onto the foreshore, then it may be decided to construct a reservoir to store the sewage for those periods when discharge would be undesirable so that it may be released at a more suitable time.

COASTAL AND OTHER SPECIAL PROBLEMS

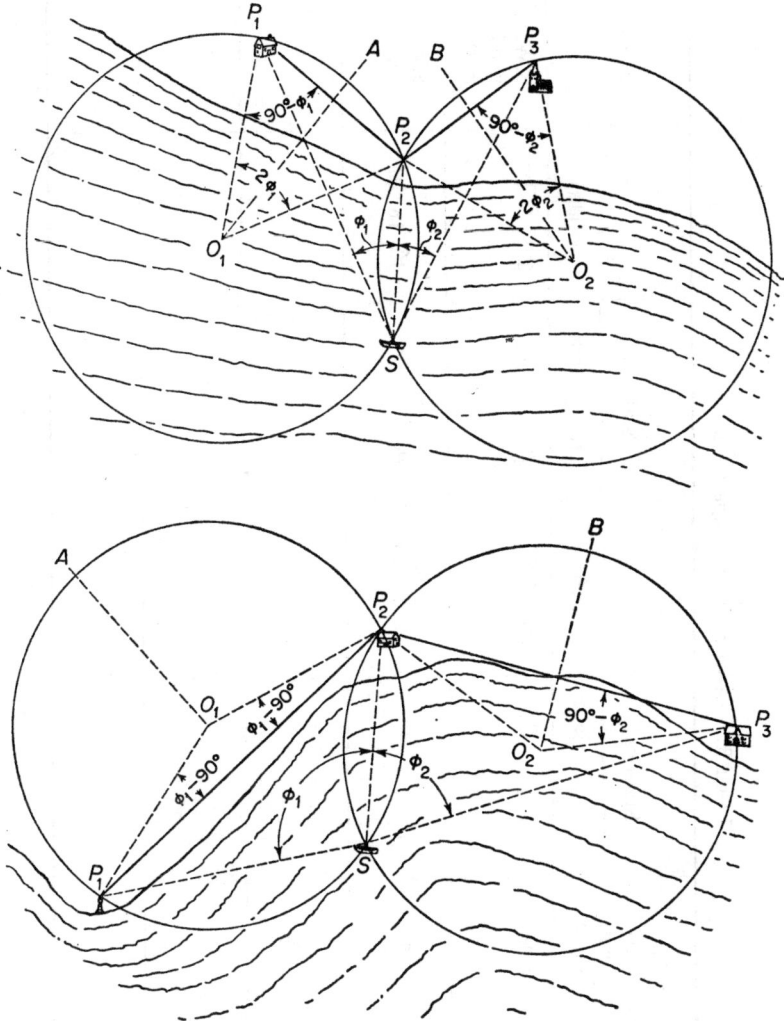

Fig. 3 Three-point problem

This control of discharge is usually by electrically-operated penstocks which are opened and closed by a 'lunar clock', that is a clock or time switch so designed that it opens the penstocks and closes them at exact times after high tide from day to day.

TABLE 31. *Booking for tidal experiments*

Date	Time	Minutes after high water	Stations		ϕ_1 (left)	Stations		ϕ_2 (right)	Remarks
			left	right		left	right		
5th April 1939	p.m. 2·45	0.00	P_1	P_2	30° 15′ 30″	P_2	P_3	27° 5′ 18″	S.W. wind fresh
5th April 1939	3·00	15	P_1	P_2	39° 16′ 16″	P_2	P_3	40° 18′ 20″	S.W. wind fresh

It will be obvious that where storage of sewage is concerned, it is best for the sewerage to be on the separate system, for it may be impracticable to store the vast quantities of storm-water that could result from heavy rainfall and, if no allowance for storing storm water were made, spill-over could be too frequent to be acceptable. But the difficulty usually is that new sea outfalls and storage tanks are wanted for towns that have existing combined or partially-separate systems. It is in such circumstances that consideration should be given to the possibility of treating the sewage.

Storage tanks

A storage tank for separate soil sewage should have a capacity sufficient to hold all the sewage likely to come down in the longest period of time during which discharge may not be made. As the tide cycle does not keep pace with the cycle of day and night, the tank will at times have to store the peak rate of flow and therefore, should it be necessary to store for 8 hours, the capacity might well have to be the equivalent of a day's flow, on the assumption that half a day's dry-weather flow can come down in 8 hours and that this figure can be doubled in wet weather. More exact capacities can be found by autographic gauging of the flow on the main sewer and allowing for the maximum condition.

If the sewerage is on the combined or partially-separate system, the storage tank becomes a storm tank, giving partial treatment by sedimentation, notwithstanding that the settled sludge may be later discharged with the sewage. (It is, of course, better if the settled sludge is kept separate and delivered elsewhere for treatment or drying.) The tank must, therefore, spill over settled storm water when it becomes full, but it does not need to have more capacity than a tank for separate soil sewage, or the required storm-tank capacity, whichever be the larger. (For partial treatment and the design of storm tanks see Chapters 10 and 11.)

Size of outfall

Sewage may be discharged to the sea by pumping or gravity. If the levels necessitate pumping, calculating the size of outfall is simple, for it is necessary only to install a pumping station capable of discharging the contents of the tank plus the incoming flow of sewage in the time during which discharge to the sea is permissible. The outfall then has to be of the economic diameter a rising main should have, which may be calculated in the manner described at the end of Chapter 3.

The problem is more complicated when discharge is by gravity, for sea outfalls are expensive structures: they have to be made large enough to discharge the requisite quantity but must not be excessively large, and consequently diameter must be determined with reasonable accuracy. Just how difficult this will be depends on circumstances.

The simplest case is when a vertical sided tank has to be emptied with free outlet or into water of constant level. Then the following formula gives a discharge coefficient:

$$C = \frac{2A\sqrt{L}(\sqrt{H_1}-\sqrt{H_2})}{t} \tag{33}$$

where D = diameter of outfall in millimetres,
  C = $0.00002078\ D^{21/8}$,
  L = length of outfall in metres,
  H_1 = maximum effective head from top water level in tank to level of water into which outfall discharges (or crown of outfall at outlet end if not submerged) in metres,
  H_2 = minimum effective head from lowest water level in tank to level of water into which outfall discharges (or crown of outfall at outlet end if not submerged) in metres,
  A = surface area of sewage in tank in square metres,
  t = time in which tank must be emptied in minutes,

and the required diameter of outfall can be found by reference to Table 32.

If, as is usually the case, there is an inflow during outflow the formula becomes

$$C = \frac{2A\sqrt{L}(\sqrt{H_1}-\sqrt{H_2})}{t} + \left\{ \frac{2ALQ}{Ct} \times 2 \cdot 3026 \log_{10}\left(\frac{\sqrt{H_1}-(Q\sqrt{L}/C)}{\sqrt{H_2}-(Q\sqrt{L}/C)}\right) \right\}$$

$$\tag{34}$$

where the additional factor Q is the rate of flow of sewage entering tank in cubic metres per minute.

COASTAL AND OTHER SPECIAL PROBLEMS 107

The disadvantage of this formula is that it cannot be solved directly: a number of trials have to be made. The amount of work can be reduced by the approximate formula

$$C = 2\sqrt{L}(\sqrt{H_1}-\sqrt{H_2})[(A/t)+Q/(H_1-H_2)] \qquad (35)$$

TABLE 32. *Discharge coefficients for sea outfalls*

Diameter (nominal millimetres)	C	Diameter (nominal millimetres)	C
100	3·86	1575	5360
150	11·17	1650	6050
225	32·4	1725	6800
300	69·0	1800	7620
375	124	1875	8470
450	200	1950	9380
525	295	2025	10 400
600	427	2100	11 400
675	579	2175	12 500
750	762	2250	13 700
825	983	2325	14 900
900	1230	2400	16 200
975	1520	2475	17 500
1050	1850	2550	19 000
1125	2220	2625	20 400
1200	2630	2700	22 100
1275	3070	2775	23 700
1350	3580	2850	25 400
1420	4120	2925	27 200
1500	4720	3000	29 000

The coefficient C, divided by the square root of the inclination (or length divided by fall), gives the discharge in cubic feet per minute according to Scobey's formula. (For square roots see Table 33.) When two or more pipes are laid in parallel under the same conditions of head, the values of C can be added together to give the value of C for one pipe that will discharge the same quantity. To convert to values for actual millimetres diameter, divide C by 1·04252.

When discharge is into the sea and accuracy is desired, allowance should be made for the fact that sea water is appreciably denser than

TABLE 33. *Square roots of numbers*

Number	Square root	Number	Square root	Number	Square root	Number	Square root
1	1·000	41	6·403	81	9·000	205	14·318
2	1·414	42	6·481	82	9·055	210	14·491
3	1·732	43	6·557	83	9·110	215	14·663
4	2·000	44	6·633	84	9·165	220	14·832
5	2·236	45	6·708	85	9·220	225	15·000
6	2·449	46	6·782	86	9·274	230	15·165
7	2·646	47	6·856	87	9·327	235	15·330
8	2·828	48	6·928	88	9·380	240	15·492
9	3·000	49	7·000	89	9·434	245	15·652
10	3·162	50	7·071	90	9·487	250	15·811
11	3·317	51	7·141	91	9·539	255	15·969
12	3·464	52	7·211	92	9·592	260	16·125
13	3·606	53	7·280	93	9·644	265	16·279
14	3·742	54	7·348	94	9·695	270	16·432
15	3·873	55	7·416	95	9·747	275	16·583
16	4·000	56	7·483	96	9·798	280	16·733
17	4·123	57	7·550	97	9·849	285	16·882
18	4·243	58	7·615	98	9·899	290	17·029
19	4·359	59	7·681	99	9·950	295	17·176
20	4·472	60	7·746	100	10·000	300	17·321

21	4·583			105	10·247	305	17·464
22	4·690			110	10·488	310	17·607
23	4·796			115	10·724	315	17·748
24	4·890			120	10·954	320	17·889
25	5·000			125	11·180	325	18·028
26	5·099	61	7·810	130	11·402	330	18·166
27	5·196	62	7·874	135	11·619	335	18·303
28	5·292	63	7·937	140	11·832	340	18·439
29	5·385	64	8·000	145	12·042	345	18·574
30	5·477	65	8·062	150	12·247	350	18·708
31	5·568	66	8·124	155	12·450	355	18·841
32	5·657	67	8·185	160	12·649	360	18·974
33	5·745	68	8·246	165	12·845	365	19·105
34	5·831	69	8·307	170	13·038	370	19·235
35	5·916	70	8·367	175	13·229	375	19·365
36	6·000	71	8·426	180	13·416	380	19·494
37	6·083	72	8·485	185	13·601	385	19·621
38	6·164	73	8·544	190	13·784	390	19·748
39	6·245	74	8·602	195	13·964	395	19·875
40	6·325	75	8·660	200	14·142	400	20·000
		76	8·718				
		77	8·775				
		78	8·832				
		79	8·888				
		80	8·944				

TABLE 33 (*continued*)

Number	Square root	Number	Square root	Number	Square root	Number	Square root
405	20·125	710	26·646	1550	39·370	3550	59·582
410	20·248	720	26·833	1600	40·000	3600	60·000
415	20·372	730	27·019	1650	40·620	3650	60·415
420	20·494	740	27·203	1700	41·231	3700	60·828
425	20·616	750	27·386	1750	41·833	3750	61·237
430	20·736	760	27·568	1800	42·426	3800	61·644
435	20·857	770	27·749	1850	43·012	3850	62·048
440	20·976	780	27·928	1900	43·589	3900	62·450
445	21·095	790	28·107	1950	44·159	3950	62·849
450	21·213	800	28·284	2000	44·721	4000	63·246
455	21·331	810	28·460	2050	45·277	4050	63·640
460	21·448	820	28·636	2100	45·826	4100	64·031
465	21·564	830	28·810	2150	46·368	4150	64·420
470	21·679	840	28·983	2200	46·904	4200	64·807
475	21·794	850	29·155	2250	47·434	4250	65·192
480	21·909	860	29·326	2300	47·958	4300	65·574
485	22·023	870	29·496	2350	48·477	4350	65·955
490	22·136	880	29·665	2400	48·990	4400	66·332
495	22·249	890	29·833	2450	49·497	4450	66·708
500	22·361	900	30·000	2500	50·000	4500	67·082

510	22·583	910	30·166	2550	50·498	4550	67·454
520	22·804	920	30·332	2600	50·990	4600	67·823
530	23·022	930	30·496	2650	51·478	4650	68·191
540	23·238	940	30·659	2700	51·962	4700	68·557
550	23·452	950	30·822	2750	52·440	4750	68·920
560	23·664	960	30·984	2800	52·915	4800	69·282
570	23·875	970	31·145	2850	53·385	4850	69·642
580	24·083	980	31·305	2900	53·852	4900	70·000
590	24·290	990	31·464	2950	54·314	4950	70·356
600	24·495	1000	31·623	3000	54·772	5000	70·711
610	24·698	1050	32·404	3050	55·227	5100	71·414
620	24·900	1100	33·166	3100	55·678	5200	72·111
630	25·100	1150	33·912	3150	56·124	5300	72·801
640	25·298	1200	34·641	3200	56·569	5400	73·485
650	25·495	1250	35·355	3250	57·009	5500	74·162
660	25·690	1300	36·056	3300	57·446	5600	74·833
670	25·884	1350	36·742	3350	57·879	5700	75·498
680	26·077	1400	37·417	3400	58·310	5800	76·158
690	26·268	1450	38·079	3450	58·737	5900	76·811
700	26·458	1500	38·730	3500	59·161	6000	77·460

sewage and that the effective head is less than the vertical distance from water level in tank to sea level. This is expressed by the formula

$$\text{Effective head} = H - (h/36) \tag{36}$$

where H = sewage level in tide-flap chamber above sea level in metres

h = height of sea level above centre of bottom end of outfall in metres

This adjustment will apply to the foregoing formulae for determining coefficient C whenever the outlet end of the outfall is submerged. Allowance should also be made for loss of head through the tide flap: this will be approximately as given in Table 17.

When, as can often be the case, discharge is into tidal waters and the tide rises above the outlet of the outfall, the problem is so complex that a mathematical solution of practical value has not been found. The simplest method then is to find provisionally the size of outfall required on the assumption of average water level in tank and average sea level during time of discharge and, with this diameter, make a stage-by-stage calculation. First, assuming that the tank is full and the tide is at the level that it will be at the moment that discharge commences, the quantity discharged during the first 15 minutes less the inflow during the same period is calculated, and from this a new level in tank is found. With this new level in tank and the new sea level after 15 minutes, the calculation is repeated to find a third level in the tank. The process is continued until the tank is empty and, if this is found to be not later than 15 minutes before the end of the permitted discharge time, the outfall may be considered of adequate diameter. If the outfall is not large enough or obviously far too large, the calculation must be repeated until the correct diameter has been determined.

The spring and neap tides for various localities in the British Isles are given in the Admiralty Tide Tables. The average curve between high and low water approximates to a sine curve and, in the absence of more accurate data, may be drawn as a sine curve for determining the various levels in the foregoing calculation. But in some places the tide varies very much from this average or there may be what is described as a double tide, and when this is the case, an autographic tide recorder should be fixed in a suitable position to record the tide curves for a sufficiently long time for average curves to be prepared for neap and spring tides and for maximum high-tide conditions.

Construction of storage tanks

Storage tanks for sewage are often in the form of tank-sewers, that is, large-diameter sewers capable of storing sewage for the requisite number of hours and emptying completely without remaining foul. The incoming sewer must have a soffit level not lower than the soffit level of the tank-sewer and, from its invert, should be a backdrop or tumbling bay taking the flow to the invert of the tank-sewer. The tank-sewer should be laid at such a gradient that, in spite of its large size, it will be self-cleansing at the flow capacity of the incoming sewer. (See Table 15 for proportional velocities of sewers flowing partly full.) The invert of the tank-sewer at its lowest level should be high enough to produce the head necessary for complete emptying of the sewer through the outfall; it will therefore be above the crown of the top end of the outfall. A tide flap should be placed between the tank-sewer and the outfall to prevent back flow from the sea and, upstream of this, should be the automatic electric penstock controlling discharge. There should be a storm- or emergency-overflow weir either set above highest tide level (if possible) or protected by a tide flap. All tank-sewers should be well ventilated and provided with access for maintenance purposes.

Sometimes, in lieu of a tank-sewer, a storage tank, rectangular on plan, is built on the sea front. This also needs to be self-cleansing and, to secure this end, it can best be designed in the form of a tank-sewer folded back on itself so that the flow through it is rapid at the end of discharge. Alternatively, it can be a mechanically-cleansed settlement tank. It is best to have two or more tanks so that one can be laid off temporarily for maintenance purposes. Such tanks should be disguised as car parks or other facilities for the public because, should visitors suspect their purpose, they are almost certain to imagine that foul odours emanate from them. Ventilating columns can easily be disguised as ornamental flagstaffs.

It is important for storage tanks on the sea coast to be constructed to resist damage by wave action. They should be protected by a sea wall that is not part of the structure of the tank, and full advantage should be taken of local knowledge as to what is required. In this connexion the maximum height of wave may be estimated by the formula

$$H = 0.32\sqrt{L} \qquad (37)$$

where H = wave height in metres,

L = width across sea in kilometres (or distance from coast to another coast line) e.g. distance across English Channel or North Sea.

Construction of sea outfalls

The type of construction required for a sea outfall depends very much on local conditions, which therefore need to be carefully studied, the type and state of repair of existing outfalls being particularly observed. In some places that are sheltered from storms or severe tidal currents, it may suffice to float out and sink a steel or plastic pipe that has been previously assembled on the foreshore. But where the action of the sea is severe, the outfall may have to be secured in position by piles and frames, in which case cast-iron pipes with socketed joints are usual. On a rocky foreshore the outfall may have to be placed in a trench filled with concrete or, above shore level, held down by straps and bolts grouted deep into the rock. Where cast-iron outfalls have to be laid under water by divers, special pipes can be made with turned-and-bored socketed joints, the turning and boring being to radius to permit some irregularity of line. The joints should each be provided with three lugs so that the spigots can be drawn into the sockets by bolts, one at the top and two below the centre line but not so near the invert that tightening-up of the bolts will be difficult.

While a tide flap is always installed at the top end of the outfall, it would be most undesirable at the bottom end unless circumstances were unusual and special precautions were taken. If there is a tide flap at the bottom end and the outfall is not held down and is empty when the tide rises, it is almost certain to float and break up.

Inverted siphons

The coast cuts across hills and valleys in a haphazard manner, interfering with natural drainage and sewerage so that pumping stations, and also means of crossing valleys, such as inverted siphons and sewers supported above ground level, are more common near the sea than elsewhere. The inverted siphon (fortunately not often required) can be an unsatisfactory device, for it needs only a modicum of bad design to make it troublesome.

The inverted siphon is a pipe which is lower at some point between the ends than it is at the lower end. This means that the pipe must always run full, because it cannot drain out, and solids that are too heavy to remain in suspension have to slide uphill under the force of the current. It follows that if the velocity in the siphon is not self-cleansing for some considerable period every day, silting must take place, reducing the cross-sectional area of flow and making the invert rough, encouraging stoppages.

The most common fault in inverted siphons is for them to be made to too large a diameter, causing the velocity to be too low at the peak rate of flow. But a siphon must be capable of taking the maximum rate of flow which, in the case of a combined or partially-separate system, may be considerably more than the daily peak. From these requirements develop the various types of inverted siphon: the simple single-pipe siphon, the single-pipe siphon with flushing tank and the multi-pipe siphon.

Simple single pipe siphon

A simple inverted siphon is a single pipe designed to have a self-cleansing velocity for several hours a day during the actual (not future estimated) peak daily flow and, under the available head, to be capable of discharging the maximum possible flow at a higher velocity. An example is a pipe which will discharge twice dry-weather flow at a velocity of 0·75 metre per second and, as might be required in the case of a separate soil sewer with a moderate degree of infiltration, a peak wet-weather flow of six times dry-weather flow at a velocity of 2·25 metres per second. The fall from the soffit of the inverted siphon at the top end to that at the lower end would have to be sufficient to induce the maximum wet-weather flow. This limits the use of this type of siphon to those cases where there is sufficient fall and the ratio of maximum flow to peak daily flow is not too great.

Siphon with flushing tank

Where there is sufficient fall available for the construction of a flushing tank operated by a true siphon, but the conditions are such that a simple siphon would not be satisfactory, it may be possible for the incoming sewage to be discharged into a flushing tank which, when full, will discharge via a flushing siphon to the inverted siphon at a rate sufficient to ensure a self-cleansing velocity whenever the siphon functions. This method has its attractions but is not always easy to apply, for the flushing tank must have sufficient capacity relative to that of the inverted siphon to give a good flush, but this capacity must be in the depth limited by the available head. The danger is that the flushing tank may become an unwanted sedimentation tank incapable of eliminating its suspended solids adequately. For this reason the method is seldom used.

Multi-pipe siphon

In most instances where a simple siphon will not serve, multi-pipe siphons are used. The arrangement is to have one pipe designed in the manner of a simple siphon to take a reasonable excess on the peak daily flow, plus one or more larger pipes to take the storm flow. The simple siphon functions most of the time, and when it is beaten by the flow, the storm water passes over a weir or weirs to the other siphon or siphons. All the siphons have inlet and outlet penstocks so that they may be laid off for maintenance purposes, individually flushed or, if of suitable diameter, used in turn as dry-weather siphons.

The following details need to be considered if trouble is to be avoided. In calculating the size of a storm-water siphon the actual depth of sewage in incoming and outgoing sewers, and not the differences of crown level, must be used in determining the hydraulic gradients. The inlet ends of the siphons must be so arranged that the crowns are slightly above the flow, permitting the entry of floating solids which would otherwise tend to accumulate in the manhole at that end.

Unless it is virtually impossible or unduly expensive, the lowest point on a siphon should be so located that a wash-out can be provided and, unless impossible, hatch boxes should be inserted on long siphons at intervals of about 100 metres. Unless there are reasons to the contrary, ventilating columns or ventilating manhole covers should be provided at both inlet and outlet ends of siphons because siphons interfere with the natural ventilation of the sewer.

Sewers above ground level

On very rare occasions sewers are carried across valleys above ground level. Hydraulically, this is a better method of traversing low points than using inverted siphons, but objections can be raised to above-ground sewers on account of their appearance or of their interference with land development.

Above-ground sewers are constructed of cast-iron or steel pipes, and may be supported on brick or concrete piers or steel structures. If the height above ground level is not very great, cast-iron pipes with a brick pier behind each socket will serve very well. For greater heights, longer spans are more economical and it is then that steel pipes may be preferred. These should be wrapped in bitumen and glass wool or asbestos fibre to protect them from the weather, but internal protection from sewage is not required.

Pipes above ground level should be laid in straight lines as far as is practicable and, if there has to be an angle in the line, provision should be made to take the thrust due to expansion. The amount of expansion allowed for is about 10 millimetres for every 10 metres length of pipe. This can be taken-up in special expansion joints at suitable intervals. But it should be borne in mind that if the pipe above ground level is a rising main under pressure, there must be no circumstance in which a joint can be blown open. Also each pipe should be well anchored half-way between each expansion joint to prevent progressive creep.

It is always advisable to provide collars of spiked railings or similar devices at each end of a line of pipe above ground level to keep children from running into danger.

PART II

SEWAGE WORKS

8
The Nature of Sewage and its Measurement

Sewage is the content of, or effluent from, sewers. It may be pure water or it may be very foul. Usually it consists of more than 99·9% of water, yet a Court of Appeal has decided that it is not water. Surface water (which as the content of surface-water sewers is storm sewage) is the run-off of rainfall and contains no pollution other than that which it may have picked up in its passage over roofs and roads. Sullage is domestic waste from lavatory basins, sinks, bidets and baths. Soil sewage contains night soil, that is the discharge from water-closets, urinals and slop sinks, and it is any mixture of soil sewage with trade waste, sullage or surface water.

It is a popular misconception that sewage mainly consists of the products of the water-closet diluted to some extent by other dirty water and that, consequently, it should make a good manure. This view is far from true as can be shown by the following figures.

The average amount of faecal matter has been stated to be 135 grammes per person per day at 75% moisture content, and the average quantity of urine 1500 grammes of water per day containing 72 grammes of solid matter. This gives a total of about 106 grammes of solid matter partly in suspension and partly in solution per person per day. But the average quantity of solids received at sewage works is about 190 grammes per head per day so that the night soil amounts to little over one-half of the total. The rest is made up of the hardness of the water supply, mineral and organic matter in kitchen waste etc., mineral and a little organic matter which finds its way in from roof and road surfaces and various industrial effluents. The last are important, for the pollution load on sewage works serving an industrial town is much greater than that on works serving a purely residential area. *More than a quarter of the organic content is oil or fat.*

The strength of sewage, i.e. the degree to which it is polluted, depends on the ratio of the organic load to the amount of clean water in which that load is carried. The quantity of water used per head per day and the infiltration of ground water to the sewers vary considerably from one locality to another and to a greater extent than the pollution load. These factors are expressed in the measured quantity of sewage and the sewage analysis, and on them are based the designs of sewage-treatment works.

Chemical tests

The main tests are for nitrogen content in terms of ammoniacal, albuminoid or organic nitrogen, nitrite and nitrate; chlorides expressed in terms of chlorion; alkalinity or acidity; oxygen absorbed from acid permanganate; biochemical oxygen demand; total solids and suspended solids. Temperatures are recorded in degrees centigrade. These tests give information not only on the degree of pollution or strength of sewage but also the extent to which changes have taken place, in particular the oxidation of the organic content.

During the present century there have been several changes in the way the results of analyses are expressed, that at present favoured being in milligrammes per litre, which is the same as grammes per cubic metre and, near enough for practical purposes, the same as parts per million. Broadly the same methods used for testing sewages are applied to partly- and finally-treated effluents and, with some additional tests, to sewage sludge.

The test for albuminoid ammonia indicates the amount of un-decomposed protein in the sewage: that for ammoniacal nitrogen (free and saline ammonia) shows that some decomposition has taken place: nitrites and nitrates are most noticeable in a treated effluent of good quality, for they show that most of the nitrogenous material has been completely decomposed to nitrogen salts.

Alkalinity can be expressed in terms of calcium carbonate and acidity can be expressed as 'acidity' equivalent of calcium carbonate. But alkalinity and acidity can be and often are expressed as pH value. This unit is not in milligrammes per litre but is a measure of the hydrogen-ion concentration, being the logarithm to the base 10 of the reciprocal of the hydrogen-ion activity; hence the alkalinity or acidity of a solution is expressed on a scale of numbers from 0 for a solution containing 1 gramme-ion of hydrogen ions per litre which corresponds to extreme acidity, to 14 for a solution containing 1 gramme-ion of hydroxyl ions per litre; and the value of 7 denotes a neutral reaction. Because the scale is logarithmic it is necessary to use logarithms if the average of a number of results is to be found, the method being to find the anti-logarithm of each test figure, average the anti-logarithms and find the logarithm. Domestic sewage is usually slightly alkaline but septicity turns it acid. A healthily digesting sludge also should be alkaline.

The amount of oxygen absorbed from acid permanganate is a quick test for finding the amount of organic carbon in the sewage. The usual test at sewage works is the 4-hours test, but there is a 3-minutes test which may be of use in the field.

The most important test for organic pollution is that for biochemical oxygen demand, which is usually the amount of oxygen absorbed by a sample incubated at a controlled temperature for a period of 5 days and known as 5 days biochemical oxygen demand or, briefly, $B.O.D._5$. This is used to measure not only the organic pollution of a sewage, but also the quality of a final effluent.

Suspended solids are measured by filtering and then weighing the dried residue, while dissolved solids are measured by weighing what is left after evaporating the filtrate: together they make the total solids content. From the suspended-solids figure can be made an estimate of the amount of sludge to be expected.

Gases in solution, for example, oxygen in solution which should always be present in a well-treated effluent, are also measured.

All sewage contains common salt, the quantity being more or less in direct proportion to the population. As sewage treatment has no effect on chloride content, a change in this between crude sewage and what is supposed to be the same sewage partly or completely treated could make the chemist suspect an accident with the sample or some dilution or other falsification of the results at the works.

In Great Britain a measure of sewage strength recommended by the Royal Commission on Sewage Disposal [21] is still used for some purposes, e.g. calculating charges for trade effluents. This is strength (McGowan) which, as quoted from the Ministry of Housing and Local Government *Methods of Chemical Analysis as Applied to Sewage and Trade Effluents*, 1956 [18], gives the strength of sewage as follows:

'Strength of sewage = $4.5 \times$ (ammoniacal nitrogen + organic nitrogen) in parts per $100\,000 + 6.5 \times$ permanganate value using N/8 permanganate in parts per $100\,000$. (38)

'The factor of 4·5 represents the amount of oxygen required to convert the nitrogen to nitrate and is correct provided that all the nitrogen is originally present as ammonia or one of its substitution products. The factor 6·5, however, is empirical and depends upon the proportion of the oxidisable carbonaceous matter which reacts with permanganate under stated conditions, and this proportion varies very widely with different trade wastes. The McGowan formula suffers from the disadvantage ... that it is based solely on oxygen requirements, although it has proved useful in the absence of appreciable concentrations of trade wastes.

'It should be noted that McGowan strength is calculated from the permanganate value using N/8 permanganate and not N/80 permanganate.... The relation between the two is no doubt variable. A figure of 1·6 has been given as the general ratio between the N/8 test carried out at 25·7° C. and the N/80 test carried out at 18·3° C. The ratio must be smaller when the M/80 test is carried out at 27° C.... and probably does not exceed 1·2. This figure could be used to obtain an approximate figure for the McGowan strength, but analysts still using it are recommended to adhere to the original method of determination....'

Sampling

Accurate sewage sampling is so difficult that if a sample is taken, divided into two parts and sent to two different laboratories for B.O.D. tests, the results may be widely divergent according to various factors, including the way the sample was handled during transit. A sample taken in a vessel that has been scrupulously cleaned may have a noticeably lower B.O.D. value than one taken in a vessel that has been merely swilled out with tap water after previous use. Samples taken by different designs of automatic sampling machine can vary wildly, and poorly designed machines will give widely different results from day to day.

It is for these reasons that opinions and calculations should not be based on individual samples but, whenever practicable, on averages taken over several weeks; hand sampling should be done with precautions and automatic samplers should be all of the same design throughout the works, of a design that is known to work satisfactorily and should be properly used.

When it is unavoidable that decisions must be made on the basis of samples taken on one day only, the recognized procedure is to take 'weighted' samples at regular intervals throughout the 24 hours. The flow is measured by weir or instrument, and a sample is taken, say, every half-hour. A quantity of each sample is measured in proportion to the rate of flow at the time it was taken, and tipped into a vessel that is stored in ice during the whole of the sampling period. At the end of the 24 hours the contents of the vessel are carefully mixed and transferred to half Winchester quart bottles. There must be no agitation of the samples to an extent which could cause aeration, and no bubbles should be included in the bottles, which should be stoppered and transported on ice to be analysed as soon as practicable, and certainly not later than 24 hours after collection.

Recently, several automatic sampling devices have been marketed or made by sewerage authorities to their own designs. These are of two main types, those that take samples into a number of separate bottles so that the change of strength of sewage throughout the day can be determined, and those that take daily average weighted samples. The latter type is of more general use and more easy to design to give representative samples.

The easiest way to take a weighted sample by machine is not to change the size of the sample according to the rate of flow (although this is done by one design) but, by an impulse from the works flow recorder, to take samples at varying intervals of time whenever some predetermined quantity (say 1000 cubic metres more or less, according to the size of the works) is recorded as having passed through the works. The samples should be drawn from a turbulent pipe or channel so that they will not contain a disproportionate quantity of scum or sludge, pumped to waste for a short time to clear the pipework of any stale sewage or sediment, after which a measured quantity should be discharged into the sample vessel. Wherever practicable this vessel should be in a refrigerator and maintained at a temperature a little above freezing point.

An automatic weighted sampler designed by the author involves no moving parts except a suitable pump. The sampler works on the leap-weir principle. Every time X cubic metres is clocked on the main instrument panel, an electric impulse sets the pump running and, by means of a time mechanism, sewage or effluent is pumped for long enough to clear the pipework. The pumping rate and the sizing of the gunmetal nozzle inside the sampler ensure that the velocity of injection is enough to throw the flow over a dam (not shown in Fig. 4) and down the waste pipe. Should the flow exceed the discharge of the nozzle, overflow takes place via the overflow pipe at the top between the rising main and the waste pipe; the vertical air vent at the top prevents siphonage. On cessation of pumping the velocity falls and the contents of the Z-shaped pipe connecting to the nozzle falls into the sample bottle (see Fig. 4).

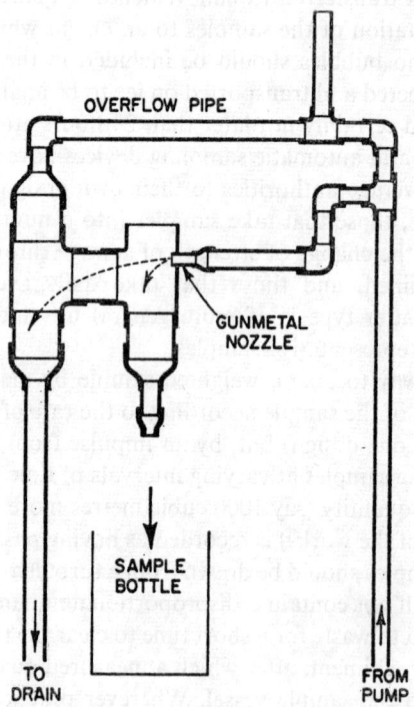

Fig. 4 Automatic sampler

The requirements of the design of this sampler are as follows:
1. The rising main must be of a suitable diameter to be free from chokage with the kind of liquid concerned, e.g. 30 millimetres, for final effluent or settled sewage.
2. The pump must have a delivery that will ensure a self-cleansing velocity in the rising main.
3. The nozzle must have a diameter that, under the maximum head

THE NATURE OF SEWAGE AND ITS MEASUREMENT

permitted by the overflow pipe, will throw the liquid over the dam into the waste pipe.

4. The Z-shaped sampling pipe must have a capacity equal to the required individual sample, between the nozzle and the point on its vertical component at which the surface of the water will stand when the trajectory of the jet throws the liquid just inside the dam and flow starts to fall into the sample bottle.
5. The vessel receiving the samples should be large enough to receive all the samples taken on the day of maximum rate of flow.

Strength of sewage

The Royal Commission on Sewage Disposal gave as an example of average sewage the following values:

	Milligrammes per litre
Ammoniacal nitrogen	35·3
Albuminoid nitrogen	9·1
Total organic nitrogen	22·5
Oxidised nitrogen	trace
Total nitrogen	58·5
Oxygen absorbed at 27°C. in 4 hours	112·7
Chlorine	91·6
Suspended solids	294·0

The strength (McGowan) was about 100 parts per 100 000 which is approximately the equivalent of a $B.O.D._5$ of 375 milligrammes per litre. The flow of sewage was 0·162 cubic metre per head of population per day.

As compared with these figures, those for the Crossness Sewage-treatment Works *circa* 1960, which works served a very large representative area in South London, were:

	Milligrammes per litre
$B.O.D._5$ (dry-weather flow)	307
$B.O.D._5$ (average flow)	282
Suspended solids (average flow)	329
$B.O.D._5$ settled sewage (dry-weather flow)	199
$B.O.D._5$ settled sewage (average flow)	186

Average daily flow 0·273 cubic metre per head of population per day.

There are very great differences in the strength of sewage at different works depending largely on the amount of water in which the polluting substances are diluted and, as mentioned above, according to the amount of industry in the district.

For the design of works it is useful to express the strength of sewage in kilogrammes $B.O.D._5$ per day, for this is a unit on which percolating filters and activated-sludge aeration plant can be designed. In the absence of analyses, an estimate of the $B.O.D._5$ load can be made on the basis of population. The following figures are typical:

	Kilogramme $B.O.D._5$ per head per day
Purely domestic sewage from a modern residential area	0·045
General design figure for a mainly domestic area on separate sewerage	0·055
General design figure for a mainly domestic area on combined sewerage	0·066
Typical balanced domestic and industrial area on combined sewerage	0·077

Standards of effluents

When sewage is discharged into a stream it decomposes, absorbing oxygen, and if the body of water and its turbulence are not great enough to cause oxygen to be dissolved from the air as rapidly as it is depleted, conditions will become anaerobic, aquatic life will be extinguished and the water turn offensive. The Royal Commission on Sewage Disposal expressed the opinion that crude sewage could be discharged and all chemical tests eliminated should the dilution exceed 500 volumes of water to one of sewage [22]. Such a condition is too rare in Great Britain to deserve consideration, and all sewage should be treated to some extent before being passed to any natural or artificial inland water-way.

The Royal Commission recommended that a final treated effluent would probably prove satisfactory if it did not contain more than 30 milligrammes per litre of suspended solids and did not take up, at 18·3°C, more than 20 milligrammes per litre* of dissolved oxygen in 5 days, provided that it was diluted by not less than eight volumes of

* The original recommendations were given in parts per 100 000.

river water. This standard was used, generally without regard to the degree of dilution, for about half a century. But recently river authorities have been calling for higher standards and to such an extent that the Ministry of Housing and Local Government found it necessary to issue a circular on the subject. The Ministry considered that, if standards more restrictive than that of the Royal Commission were demanded as a general rule, without reference to particular conditions, this could lead to diversion of resources from more important tasks. The Ministry suggested that a river authority should not require a higher standard than that of the Royal Commission without explaining their reasons in detail to the local authority or, in the event of a local investigation, to the Ministry's inspector, but where the need for a higher standard could be demonstrated, the local authority would have to pay for the cost of the works involved. It was stressed that river authorities should not demand a $B.O.D._5$ limit of less than 20 milligrammes per litre without first taking into account all the data relating to the river as regards oxygen replenishment etc., which could have a bearing on the need for a particularly high standard of treatment.

The easiest way of producing an extra-high quality effluent is not to build larger sedimentation and aeration tanks but to remove a greater proportion of suspended solids from the final effluent by filtration. This not only removes suspended solids but also makes a $B.O.D._5$ reduction equal to about one-third of the suspended solids reduction. Accordingly, the Ministry considered that, if the suspended solids standard were to be reduced from 30 to 20 milligrammes per litre, the $B.O.D._5$ value could be reduced from 20 to 17 milligrammes per litre, if the suspended solids figure were reduced to 10 milligrammes per litre, the $B.O.D._5$ could be reduced to 15 milligrammes per litre, but no significant reductions would be justifiable below this figure even if the suspended solids limit were reduced to 5 milligrammes per litre.

Some river authorities have called for minimum concentrations of nitrates, but Ministry opinion was that such a demand was not appropriate except in the cases of certain industrial rivers.

Another matter, easily overlooked, is that there is not much point in calling for an effluent of higher standard than usual when much greater pollution is being caused by discharge of crude sewage from storm overflows or, on certain days of the year, by the overflow of settled sewage from storm tanks (see Chapter 11).

Measurement of flow

Flows of 0·15 to 7 cubic metres per minute can be measured to an accuracy of $\pm 1\cdot 5\%$ with a 90° V-notch thin-plate weir for which the formula is

$$Q = 0\cdot00000291 H^{2\cdot 48}, \qquad (39)$$

where Q = discharge in cubic metres per minute,
H = The measured head over the bottom of the V in millimetres.

For larger flows a fully contracted rectangular thin-plate weir may be used and, provided the length of the sill is not less than 300 millimetres and the head over it not less than 75 millimetres, the accuracy will be $\pm 2\%$. The formula is

$$Q = 0\cdot00000345\,(L-0\cdot 1H)H^{1\cdot 5}, \qquad (40)$$

where Q = discharge in cubic metres per minute,
L = length of weir in millimetres,
H = measured head over weir in millimetres.

Thin-plate weirs should be edged with non-corrodable metal 1·5 millimetres thick, sharp on the upstream side and chamfered at 45° on the downstream side. There should be a free fall of 75 millimetres from the lowest part of the weir to the highest downstream water level, and the downstream invert should not be less than 150 millimetres below the lowest part of the weir. The sides of the channel upstream of the rectangular weir should not be nearer to the sides of the opening by less than four times the maximum head, and the invert should be three times the maximum head below the weir. The sides of the upstream channel above a V-notch weir should not be less than one and a half times the maximum head from the nearest part of the notch. The head over any thin-plate weir should be measured at a distance of six times the maximum head upstream of the weir, and the velocity of approach should not exceed 150 millimetres per second.

Thin-plate weirs are used for liquids free from sediment or for temporary recording of sewage flows. But they have disadvantages when applied to sewage and sewage-works liquids in that the low upstream velocity causes settlement of solids. This difficulty is overcome by use of the standing-wave flume which consists of a constriction in a channel. There are several types of standing-wave flume, including rectangular flumes with side contractions only, rectangular flumes with humps in the invert, rectangular flumes with both side contractions and humps, and also V-notch flumes and flumes of trapezoidal and semi-circular or parabolic cross-section.

For most purposes when the flow is more than 1 cubic metre per minute the rectangular flume is used. For this the formula is

$$Q = 0{\cdot}00000324 B H^{1 \cdot 5}, \tag{41}$$

where Q = discharge in cubic metres per minute,
B = width of opening in millimetres,
H = head over invert at point of contraction in millimetres.
(See also equation 46).

The depth H is measured from the level of the top of the hump or, in the absence of a hump, from the invert level of the throat, to the top-water level at a distance of at least $3H$ above the hump or contraction. The length of the throat should be $2H$. Except where humps only are used, the preferable value of B is approximately equal to the maximum value of H. The radius of curvature of the side contractions upstream of the throat should not be less than $2B$. When there is a hump, the upstream end of it should be formed to radius so that the curved length is equal to the length of the curve of the sides but, in the absence of side contractions, this radius should not be less than $2H$. Downstream of the contraction the flume should increase in width, each side sloping outward at about 1 in 6, and the floor should fall away to such an extent that, at minimum rate of flow, the water level in the channel below the flume is at least 25 millimetres below the invert of the throat. In order that no disturbance shall occur in the flume, there should be a length of straight channel above the throat equal to at least eight times the width of the incoming channel.

The float chamber should be connected to the main channel by a small opening measuring about 150 millimetres square. Arrangements should be made for flushing the float chamber with sewage to remove sediment. The float chamber should measure about 0·75 metre square and its invert should be set below the invert of the throat of the flume to permit proper submergence of the float.

Standing-wave flumes are used for recording most sewage-works flows such as flows of crude sewage, sludge, returned activated sludge, etc. Records are made either by means of recording instruments on local charts or by transmitting them to main instrument panels elsewhere.

Venturi meters, which are very useful for measuring flows of clean water, have limited application at sewage works. They serve very well for measuring effluent and activated sludge and can be used for sewage, with precautions. They should have water under pressure supplied to them for clearing the pipework and gauging arrangements.

Of particular use for measuring large flows of sewage or sludge is the Foxboro-Yoxall magnetic flow meter. This instrument has no pitot tubes which can be choked. It depends for its operation on the effect of the flow of electricity through the passing liquid in a magnetic field. In small sizes it is costly but it can compete in price with the larger Venturi meters. Another advantage is that it does not require a straight line of pipe for some distance upstream.

9
General Requirements of Sewage Treatment

The chief methods of removing pollution from sewage are separation of suspended solids by settlement, oxidation of the dissolved solids and any remaining suspended solids by natural decomposition with the aid of micro-organisms, and removal by further sedimentation of solids rendered settlable or produced by the aeration processes.

There are ancillary works. Sewage is screened to remove any large solids that might interfere with pumps or other parts of the works; heavy mineral detritus that could choke the sludge hoppers of sedimentation tanks or form banks in sludge tanks is removed; the sludge from primary and final sedimentation is removed from the works, usually after some form of treatment such as digestion or drying; and, where the sewerage is on the combined or partially-separate system, there are tanks for the storage or partial treatment of storm water.

Siting of works

Sewage works are usually sited below the area which they serve so that the sewage naturally gravitates to them, and they are above and near to the natural watercourse into which the treated effluent is discharged. In the event of this ideal position not being practicable, the sewage has to be pumped to a site which is not below the drainage area or, should the sewage works come at such a level that the weirs of the final sedimentation tanks are below the flood level of the river or stream, effluent has to be pumped.

There is some loss of head through sewage works which, unless plenty of natural fall is available, must be economized. A percolating-filter scheme requires a depth of not less than 2·5 metres from the soffit (or sometimes the invert) of the incoming sewer to the highest flood level of the river, and it is usual to allow a minimum of about 3·75 metres. Activated sludge works do not necessarily use up so much head, a small plant requiring about 0·5 or 0·6 metre. But at large works more loss is involved because of the greater depth over tank weirs and distances apart of the components. Also, constant-velocity detritus channels and measuring flumes for large flows may require considerable falls. The figures for the Crossness Sewage-treatment Works, which were designed to serve a population of 1 700 000, are quoted as an example: at the maximum rate of flow to be treated, the loss of head through the works is about 5 metres, of which nearly two-thirds is used in the standing-wave flumes which control the detritus channels and measure the rate of flow.

Sufficient land should always be acquired for present and any future works. On several occasions, too little reserved land has caused difficulties and expense to the sewerage authorities concerned. Once land has been acquired and reserved for sewage works, it should be used with due economy.

The amount of land required can be determined by making an estimate of the area that will be covered by the component parts of the works with their embankments, access roads and ancillary buildings, allowing for considerable future extension: then the area so found can be doubled for safety. This method has seldom proved extravagant for, over the years, standards of sewage treatment have become more stringent calling for greater areas of land than would originally have been thought necessary.

A fair estimate of the area of land to acquire is given by the following formula, which is based on the actual areas at all the twenty sewage-treatment works in Greater London which individually serve populations ranging from about 3 000 000 to 60 persons:

$$A = P^{0 \cdot 6}/74, \qquad (42)$$

where A = area of sewage works property in hectares,
P = head of population.

The above is the formula for an enveloping curve which picks up all except two of the works concerned, one of which includes land treatment. The average area is two-thirds of the figure given by the formula.

For economical development, needing a minimum of excavation, the sites of sewage works should fall towards the river at a slope of about 1 in 50 for a percolating-filter scheme, or slightly sloping for an activated-sludge plant or land treatment works. For land treatment a sandy or similar porous subsoil is best, and a heavy clay subsoil least satisfactory. But it can happen that the proximity of sewage works to a river means that they are on some depth of alluvium and in some cases peat, which may mean that the works have to be constructed on piles at an additional cost of 15% to 20%.

While the cost of sewage-treatment works can vary according to the method of treatment, the standard of provision and local problems, costs are usually so regular that estimates can be made on the basis of population served. In the year 1959 a curve of capital costs was prepared which was found to average

$$\text{Cost per head of population} = £55/P^{0\cdot 16}, \qquad (43)$$

where P = head of population served.

Allowing for change in the purchasing power of the pound, the £55 would have to be altered to about £75 to give costs for the year 1969. This cost excludes the cost of land, which can be a considerable item.

Type of works

By far the greatest number of sewage-treatment works in Great Britain are percolating-filter schemes in which the aeration of settled sewage is effected on beds of clinker, broken stone or similar material through which settled effluent is trickled. The method is applicable to works of all sizes, from those serving individual houses to comparatively large works such as those at Reading, which serve a population of 118 000. If properly designed, a percolating-filter scheme can be the least expensive in total annual charge on the rates.

The activated-sludge systems of diffused air and surface aeration are used at the very largest works, such as London's Beckton and Crossness Sewage-treatment Works which serve populations of nearly 3 000 000 and 1 700 000 respectively. The methods can be applied to very small works but, as a general rule, they should be confined to circumstances in which the processes will be under regular supervision by qualified engineers and chemists. They have the advantages of reduction of sewage-works odours and absence of the filter flies which are to be found at both percolating-filter and land-treatment works: also they require less land than do other methods. They are particularly suitable if it is necessary to have sewage works near to developed areas.

Land treatment, which was once the only method in use and which survived for many years because it could be economical where land was cheap, is being employed less and less. As the method no longer deserves a chapter to itself, it will be described here.

Land treatment should come after sedimentation for, although it is possible to treat on land sewage which has not been settled, a much greater area is then necessary. In the method known as 'land filtration' the tank effluent is irrigated into parallel trenches which are dug with level inverts throughout their lengths, so that the sewage does not run down them to the lowest point. Intermediate between these trenches, agricultural tiles are laid a moderate distance below the surface and connected to a main drain that leads to the outfall. The agricultural drains should not be laid at right angles to the irrigation ditches, as is often recommended, for this encourages short-circuiting from the bottom of the trench straight down to the

TABLE 34. *Areas of land required for land treatment*

Class of soil and subsoil, and methods of working	Cubic metres of settled sewage per hectare per day	Area required for sludge disposal per 1000 cubic metres of sewage per day (hectares)
Class I. All kinds of good soil and subsoil, e.g. sandy loam overlying gravel and sand:		
(a) Filtration with cropping	135	0·445
(b) Filtration with little cropping	280	0·445
(c) Surface irrigation with cropping	78	0·445
Class II. Heavy soil overlying clay subsoil:		
Surface irrigation with cropping	56	1·07
Class III. Stiff clayey soil overlying dense clay:		
Surface irrigation with cropping	34	1·78

GENERAL REQUIREMENTS OF SEWAGE TREATMENT 137

drain through the earth that was disturbed when the drain was laid. This has been an occasional cause of a very bad effluent.

'Broad irrigation' is used in all cases where the ground is not sufficiently porous for land filtration, and sometimes even where land filtration would be possible, because it involves less capital cost. In all circumstances, it is less efficient than land filtration and requires the use of more land. The tank effluent is distributed by a trench which is dug along the contour at the highest part of the land to be used, and provided with hand-stops or with agricultural tiles that can be plugged with clay as a means of distributing the flow onto the part of the land to be used at any one time. The ground is ploughed *in the direction of the contours* so that the tank effluent will flow from furrow to furrow, taking a long time to find its way downhill. The effluent is then collected by another ditch, to be redistributed at a lower level and collected again by a third ditch, and this process may be repeated as many times as the site permits.

For the area required for treatment of sewage on land, some guidance can be obtained from Table 34, which is based on figures in the Fifth Report of the Royal Commission on Sewage Disposal [21].

Also the writers of the General Report dealt with the question of the volume of sewage that could be treated on land as follows:

'To summarise all our results within the limits of a few sentences is impossible, but we may say in conclusion, and speaking in general terms, that we doubt whether even the most suitable kind of soil worked as a filtration farm should be called upon to treat more than 30,000 to 60,000 gallons per acre per 24 hours at a given time (750 to 1,500 persons per acre); or more than 10,000 to 20,000 gallons per acre per 24 hours, calculated on the total irrigable area (250 to 500 persons per acre). Further, that soil not well suited for purification purposes, worked as a surface irrigation or as a combined surface irrigation and filtration farm, should not be called upon to treat more than 5,000 to 10,000 gallons per acre per 24 hours at a given time (125 to 250 persons per acre); or more than 1,000 to 2,000 gallons per acre per 24 hours, calculated on the total irrigable area (25 to 50 persons per acre).

'It is doubtful if the *very worst* kinds of soil are capable of dealing satisfactorily even with this relatively small volume of sewage. The population per acre is calculated on 40 gallons of sewage per head per day. It is here assumed that the sewage is of medium strength, and is mechanically settled before going on to the land.'

According to present-day methods of reckoning it would appear that about 75 square metres of the *best* land could be allowed, per kilogramme per day of B.O.D.$_5$ in the crude sewage, for the area in use at any one time and three times this amount of land should be available. For poor quality land as much as six times the above areas could be required.

Treatment of trade wastes

Industrial wastes are discharged to the sewers by agreement between the firms concerned and the local authority, but it is often advantageous to treat or partially treat the wastes at the company's works. First, the local authority will most probably require that the wastes shall be reduced in temperature, that any acid or substance liable to damage the sewers shall be neutralized, any grease or heavy sediment removed and any substance that could produce poisonous gases in the sewers eliminated. The local authority may also require balancing tanks to regulate the discharge. Second, it can often be an advantage to the firm concerned if they partially treat their effluent by removing substances which have salvage value.

Trade wastes thus involve several processes which are particular to the wastes concerned. But often they are amenable to treatment in a similar manner to sewage by screening, sedimentation and aeration with the aid of micro-organisms. There are, however, some organic wastes, such as that of the citrus fruit industry, which cannot be aerated unless nutrients are added to assist the micro-organisms or unless they are mixed with a sufficiently large body of domestic sewage.

Charges for trade wastes discharged to the public sewers can be based on proportional costs of the component parts of the municipal sewage works and their operation. The Mogden formula, used for this purpose, is based on the annual repayment of loan plus the annual running costs divided under the heads of the works as a whole, the removal of suspended solids and aeration. This formula, converted to metric units and with the values of X, Y and Z for the figures which in the original formula applied to the cost in Middlesex in 1957, reads

GENERAL REQUIREMENTS OF SEWAGE TREATMENT 139

$$\text{£ per 1000 cubic metres} = X + \frac{S}{Y} + \frac{M}{Z}, \qquad (44)$$

where S = suspended solids in trade waste in milligrammes per litre,
M = strength (McGowan) in parts per 100 000,
X = a constant depending on the annual cost of those parts of the works that do not relate to sedimentation or aeration,
Y = a constant depending on the annual cost of those parts of the works that relate to primary sedimentation and disposal of primary sludge,
Z = a constant depending on the annual cost of those parts of the works that relate to aeration and final sedimentation.

Layout

The layout of the works should be orderly and logical. The various units should be in relative positions that will not only facilitate future extension to a preconceived plan but also will allow for any reasonable possibility. Care should be taken to economize land, particularly where the scheme is small: the designer should not spread the works out for the sake of filling up the field that has been acquired. On the other hand, there should be sufficient working space around the works as a whole and, at large works, between the various blocks of tanks, etc., to give access for the contractor during construction. Excavations and embankments should be neatly executed to a well-conceived plan in which no details are glossed over. There should be a balance of cut and fill so that neither does surplus earth have to be removed nor borrow pits dug. The plan of the works should have a pleasing, well-balanced appearance.

Pipework and channels

The various conduits at sewage works include gravitating mains, rising mains and inverted siphons, together with several open channels, either embodied in the structure of the tanks or connecting from one unit to another. In all of these the velocities should be as required to give self-cleansing conditions with the fluid concerned. Pipes, culverts and open channels for crude sewage should flow at not less than 46 metres per minute, and all rising mains except those for sludge should flow at about 50 metres per minute. Mains for settled effluent may have lower working velocities in the region of 37 metres per minute but if, as in the case of percolating-filter feed pipes, they are inverted siphons, they should have washouts for removal of any heavy solids that might inadvertently accumulate in them.

Flow of sludge in pipes

Researches in several countries have shown that the critical velocity at which laminar flow changes to turbulent flow is very high for sludge and very varied according to the nature of the sludge. It has frequently been stated that there is a lower critical velocity below which flow is laminar and sluggish, and an upper critical velocity above which the flow is fully turbulent and in accordance with flow formulae for water, while between the two is a region in which flow may be either laminar or turbulent or a combination of both conditions. All experiments with which the author has been concerned have

TABLE 35. *Approximate minimum hydraulic gradients for sludge mains*

Diameter (actual milimetres)	Critical velocity (metres per minute)	Gradient (length/fall)	Discharge (cubic metres per minute)
100	111	18	0·871
125	102	28	1·25
150	94·7	41	1·67
175	89·2	56	2·16
200	85·5	72	2·69
225	82·3	90	3·27
250	81·4	105	4·00
300	77·6	145	5·49
350	75·7	185	7·28
375	75·1	205	8·29
400	74·6	225	9·38
450	73·3*	270	11·66

* Minimum velocity for all larger diameters is about 70 metres per minute.

shown one critical velocity only for one diameter of pipe at which there is an abrupt and complete change between laminar and turbulent flow.

Table 35 is based on the average of values for the upper critical velocity for digested or semi-digested sludge as found at five overseas sewage works. This average is about 30% higher than the average for the lower critical velocity for the same works and about 15% higher than the critical velocity found in the experiments with which the author was concerned. The table should therefore be reliable for use for most sludge-flow calculations. Where, however, very extensive sludge pipelines may have to be constructed, it would be advisable for adequate experiments to be made with the sludge concerned.

The practice of some engineers is to design sludge mains so that the flow is never at less than the critical velocity: this not only simplifies calculations but ensures a self-cleansing condition.

10

Preliminary Treatment of Sewage

Preliminary treatment, which means screening and detritus removal, is that part of sewage treatment which always comes before storm-water separation. It will be appreciated, therefore, that preliminary-treatment works may have to deal with very large flows from combined or partially-separate sewerage systems, or with very small flows from separate soil sewers, and that there must be considerable differences in the type and details of the works concerned.

Screening and detritus settlement precede partial treatment in storm tanks whether or not these are located at the sewage-treatment works, or they may be the only treatment given to sewage discharged into tidal waters. They are omitted at those very small works which take purely domestic sewage from isolated buildings, such as country houses and institutions, and where sedimentation takes place in septic tanks, not the type of sedimentation tank used at municipal works.

Screening at small works

At small works serving a few thousand head of population it suffices to have hand-raked screens. The recognized rule-of-thumb is to provide a submerged screen area of 1 square metre per 7000 head of population on the assumption that the screens will be raked three times a day: with a reduced frequency of raking, the area should be increased in inverse proportion. Hand-raked screens are usually installed in a detritus tank, for the chamber occupied by the screen must reduce flow velocity and cause settlement, and as detritus removal is required, the one chamber can serve the double purpose. While the side walls of the tank must be vertical to accommodate the screens, the bottom should be in the form of a hopper or double hopper before and behind the screen, having 45° slopes and sludge valves

PRELIMINARY TREATMENT OF SEWAGE   143

or penstocks for drawing off the detritus. The screen, which should be removable, should slope at an angle for easy hand-raking up to a channel from which the screenings can be shovelled. Care should be taken in the design that sludge outlets can be easily cleared should they become blocked by heavy detritus.

It is the rule to have two detritus tanks so that one can be laid off while the other is being given any necessary maintenance. The capacity of a total of one-fiftieth dry-weather flow, recommended by the Royal Commission on Sewage Disposal [21], is no longer considered desirable, being too great and liable to cause settlement of organic material which would better be settled in the sedimentation tank. Usually it suffices to allow a top-water surface area of 1 square metre for, say, every 1555 cubic metres per day at the maximum rate of flow on the basis of reasonable settlement of all particles of siliceous material down to 0·2 millimetre diameter (see Table 36).

TABLE 36. *Settlement speeds of siliceous particles*

Diameter (millimetres)	Falling speed (millimetres per second)
0·13	10
0·20	18
0·50	50
1·00	100
4·00	250
10·00	430

Comminutors

Comminutors combine in one simple mechanism both screen and disintegrator for breaking up screenings. Although they can be used at works taking any rate of flow, they are particularly advantageous where the flow is greater than could be dealt with by the simple type of screen- and detritus-tank installation by themselves, but too small to be dealt with by vertical mechanically-raked screen.

A Comminutor consists of a vertically rotating drum of highgrade cast-iron provided with slots about 6·3 millimetres to 9·5 millimetres wide in the smaller- and larger-size machines respectively, and housed in a concrete chamber in such a manner that the sewage is brought to the outside of the drum, passes through the slots and then downwards on the inside to be discharged downstream of the chamber. The screenings intercepted by the slots are carried by the slow rotation of the drum until they are intercepted by a steel comb against which they are cut up by teeth arranged on the trailing ends of the drum slots.

The required size of Comminutor can be determined by charts provided by the manufacturer. From these it will be found that the machine can be used for works of very varied sizes up to the largest. The only limitation to the use of Comminutors is that their head loss must not be less than 50 millimetres for the smallest machine to 100 millimetres for the largest, and the maximum head should be 180 millimetres for the smallest machine and 380 millimetres for the largest. They should be so located relative to the level of the sewage that the head loss through the orifices at maximum rate of flow does not exceed the appropriate maximum head and that, at one and a half times the average daily rate of flow, it is sufficient to hold the solids against the slots during the cutting operation. A standing-wave flume should be installed downstream to adjust the depth of flow to suit the characteristics of the Comminutor.

It is preferable for the Comminutors to be placed downstream of the detritus channels because the removal of detritus reduces the wear on the cutting parts. There is no need to protect Comminutors by screens or any other device, for they are quite capable of dealing with the vast majority of floating solids that come down the sewers with less risk of damage than there would be with some types of mechanical screen.

Mechanically-raked screens

Screens with power-operated raking mechanisms are generally provided whenever flows are sufficient to maintain, in clean condition, a chamber or channel large enough to house the smallest available mechanized screen which, in the case of one make, means a channel width of 1·37 metres and a maximum working depth of 1·07 metres. There are several types of mechanical screen, including rotating perforated plates and rotating perforated or slotted drums, but apart from the Comminutor described above, most of the screens installed in Great Britain are bar screens with vertical or steeply sloping bars cleansed by rakes that carry the screenings to the surface. The particular advantage of this type of mechanism is that it can be used in chambers that are very deep below ground level, where most other designs would be impracticable.

The term 'fine' screens is applied to screens with slots or perforations of not more than 6·3 millimetres diameter or width. All bar screens fall into the classes of 'medium' screens with clear openings of 8 to 38 millimetres and 'coarse' screens with openings of more than 38 millimetres. The last are rarely used, and for most purposes screens with openings of not less than 13 or more than 19 millimetres width are to be preferred, for smaller widths can involve mechanical difficulties and those that are wider let through too great a proportion of those solids which are unsightly when they settle on the weirs of tanks and which have been known to stop centrifugal pumps.

The main difficulty in the design of chambers for mechanically-operated screens is that of avoiding settlement, for such chambers should not be detritus tanks; detritus at all except the smallest works should be settled in special tanks or channels designed for the separation of mineral grit virtually free from putrescible material. This means that the velocity through the screen chambers must be at about 0·76 metre per second. But as the screens usually have side frames occupying in the region of 0·3 metre of the width of the chamber and the bars may occupy as much as one-half of the remaining width, the velocity through the bars will be considerably greater than the velocity through the chamber; yet it must not be excessive. For many years it was British practice to have low velocities through screens, and dirty, silted screen-chambers resulted. But provided the screen is constructed to take the water pressure, a differential head of 1 metre between upstream and downstream sewage levels is permissible, and does not interfere with the operation of a satisfactory mechanism. Nevertheless the maximum velocity through the bars is usually limited to 1 metre per second. Thus, the capacity of a screen can be taken as being the flow which can pass through the spaces at the maximum submergence at this velocity, assuming that the screens are kept clean. (In practice it is usual to make allowance for the screens being partially covered with screenings without unduly backing up the incoming sewer.)

The flow through a vertical screen can be calculated by the orifice formula

$$Q = mA \sqrt{(2gH)}, \tag{45}$$

where Q = cubic metres per second,
H = head in metres from water level upstream to water level downstream,
A = submerged area of openings in screen calculated by multiplying the sum of the width of the openings in metres by the vertical depth less one-third of the head in metres (assuming that the upstream water level is not above the top of the screen),
g = acceleration under gravity, 9·80665 metres per second per second,
m = a constant for the type of orifice which in the case of clean bar-screens is about 0·6.

In some instances there may be varying submergence of screens. For example, if a large-diameter sewer enters the screen chamber the submergence can vary with the depth of flow in the sewer (see Table 15) and therefore calculations of velocities through screens and chamber should be made for various depths. Also, if downstream of the screen chamber are constant-velocity detritus channels controlled by standing-wave flumes, calculations must be made to ascertain that the screens will not be backed up on the downstream side at any rate of flow.

The bars of screens should be taper-bars about 12·7 millimetres wide on the upstream side tapered to about 0·95 millimetres on the downstream side to reduce the likelihood of solids jamming between them.

Mechanical screens are usually automatically started and stopped, the most common arrangement being differential float gear which causes the raking mechanism to run whenever the difference of head above and below the screen is more than a predetermined amount. A differential head of 150 millimetres was commonly allowed, but this is not enough if velocities through the screens in the region of 1 metre per second are allowed in design, and, therefore, the limits of adjustment should be specified.

Screen installations should always be designed on the assumption that one screen may have to be laid off for maintenance and that the remaining screen or screens will have to take the maximum rate of flow. Where there are several screens, more than one stand-by is desirable.

Disintegration

Screenings are objectionable material, unpleasant to deal with by manual labour and, as they attract flies, rats and other vermin, a danger to health. It is for these reasons that screenings are broken up by Comminutors or other disintegrators and returned to the flow of sewage at all except the smallest works where the use of such machines would be impracticable.

Perhaps the best-known machines of this kind in Great Britain are the Sulzer disintegrators. These are manufactured in various sizes and are in the nature of low-lift axial-flow pumps in which the cutting impeller cuts up the screenings against a blade upstream of a grid. The recommended working conditions are as given in Table 37. The

TABLE 37. *Recommended working conditions for disintegrators*

Size	Suction diameter (actual millimetres)	Delivery diameter (actual millimetres)	Delivery (cubic metres per minute)	Output of screenings (cubic metres per day)	Total head (metres)
D8"	203	203	1·82	5·7	1·83
D10"	254	203	2·27	7·1	1·68
D12"	305	254	4·55	14·2	2·44
D15"	381	305	7·73	25·5	2·28

manufacturers' recommended output of screenings is based on the assumption of dilution in the ratio of 100 volumes of water to one of screenings plus an allowance for variations within the daily rate.

To prevent consolidation of screenings upstream of the disintegrators, the sump from which they draw should not exceed 0·7 by 0·9 metre on plan even when two machines are to be installed (except in the case of the D15"). The bottom of the sump should be 7·5 millimetres lower than the invert of the disintegrator suctions to collect heavy metal objects that could do damage. If the diluent water is sewage, it should be drawn from downstream of the screens, the pipe being of the same bore as the suction branch of the disintegrator pump; the soffit of the pipe should be below the lowest possible water level and a screw penstock or valve should be fitted to control the flow. The soffit of the disintegrator suction should be at the level of the invert of the diluent pipe and the suction piping should be of the same bore as the disintegrator branch, as short as possible and isolated by a sluice valve. A flexible joint is required between the valve and the disintegrator to facilitate maintenance.

If the screens are operated by differential float switches, the disintegrators should start immediately the raking gear starts but should not stop until the screenings that have been brought to the surface are washed into the disintegrators and the sump left clear of solids. Also, the disintegrators should not run for such a short time that they will restart so frequently as to overheat the electric starting gear. It has been suggested that disintegrators should run continuously together with the screens, but in most circumstances to arrange for them to do so would be too wasteful of power.

Quantities and properties of screenings

The quantity of screenings varies depending, among other factors, on the size of the sewerage system. If the drainage area is small, there may be a proportionally greater amount of faecal matter, but in large areas organic material tends to be converted to sludge during transit through the sewers and the screenings, reduced in quantity, contain a greater proportion of more durable materials such as rag, plastics and rubber.

Some indication of the quantities of screenings to be expected is given in Table 38. Imhoff and Fair gave the somewhat smaller figures of 0·0155 to 0·0078 cubic metre per 1000 head of population per day for 12·7 and 25·4 millimetres respectively and, as the peak rate of collection, five times the average [12]. In London, where most of the sewage has to travel many miles before reaching the treatment works, the quantity of screenings is stated to be 0·00684 cubic metre per 1000 head of population.

TABLE 38. *Effect of screen spaces on quantity of screenings*

Clear spaces between screens (millimetres)	Cubic metres per day per 1000 head of population
12·7	0·0227
19	0·014
25·4	0·0085

Screenings have a moisture content of 80% to 85% and weigh about 700 to 900 kilogrammes per cubic metre. About 85% of the dry matter is volatile.

Another method of disposing of screenings is to burn them in gas, oil or coal furnaces at sufficiently high temperatures to prevent disagreeable odours. But the method is not popular in Great Britain because it is expensive, requiring 3700 to 5800 kilojoules per kilogramme of screenings destroyed on the basis of weight before dewatering. It is necessary to de-water the screenings by pressure before incineration to make the method at all practicable.

Detritus settlement

If detritus is settled in tanks of sufficient size to effect sedimentation of the smallest siliceous particles likely to interfere with the performance of primary sedimentation tanks or sludge digestion tanks, a heavy mixture of gravel, grit and putrescible organic matter is collected. This material is useless and liable to become offensive, and present-day practice is either to wash the settled detritus to remove the organic matter or so to design the detritus channels that comparatively clean detritus is settled, the organic matter being carried away with the flow. There are several different designs of such detritus tanks or channels.

The Dorr Detritor

The Dorr Detritor is a square tank with chamfered corners. The greater part of the floor is flat and circular, the remainder sloping towards it. It has a surface area of 1 square metre for every 1555 cubic metres per day maximum rate of flow (or such other figure as the engineer may desire according to the smallest particle to be settled) so that all particles of 0·2 millimetre diameter and upward may be capable of falling from the surface to the floor during the detention period, and is made deep enough to ensure that the maximum cross velocity does not exceed 0·3 metre per second. The inlet consists of a number of deflectors arranged across one side of the tank: the outlet is a weir across the other side. A rotating mechanism pushes the settled detritus into a sump from which it is pumped to a grit washer.

The grit washer consists of a sloping channel in which a ladder-like oscillating mechanism pushes the grit up the slope against a downward stream of wash-water. Eventually the washed grit passes from the top of the ramp into a waggon while the organic material is washed down the ramp and returned to the tank. There is some tendency for certain materials such as tea-leaves to be returned to the tank and collected again by the mechanism, accumulating in the circuit instead of passing on with the effluent. Means for removing these should be provided.

Aerated detritus channels

At works where air at suitable pressure is available, such as diffused-air activated sludge works, reasonably clean grit can be collected in detritus channels which are agitated by aeration. A line of dome diffusers or a perforated pipe is arranged near the floor along one side of a spiral-flow aeration channel, in such a position that it will not foul any dredging mechanism. The rate of air flow is adjusted until a reasonably clean grit is settled, most of the organic material being passed on with the effluent.

These channels can be of rectangular cross-section and of surface area similar to that of a Dorr Detritor installation. At the outlet end is a weir to control the depth, but the depth is not important except that the longitudinal velocity should never exceed 0·3 metre per second. Grit can be removed by travelling dredger or can fall into a grip at the diffuser side to be discharged by air-lifts. Arrangements should be made for intercepting and decanting grease at the outlet end of the channel, for aeration tends to bring grease to the surface.

Constant-velocity detritus channels

Probably the simplest and most effective means of settling detritus in comparatively clean condition is the constant-velocity detritus channel in which velocity is maintained at almost exactly 0·3048* metre per second at all rates of flow. There have been several ways in which this condition can be achieved. One is to have several channels which can be brought into service manually according to the flow at the time; this is wasteful of labour and not too effective.

A reasonably constant velocity can be maintained in a channel of rectangular cross-section by providing a standing-wave flume with

* The old rule of 1 foot per second.

the same depth and invert level as the channel, and with maximum discharge capacity equal to one-half the required flow, together with a wide, low orifice with its centre at the invert of the channel and of the area required to take the remainder of the flow. In the United States specially-shaped weir plates are used to give more accurate control of velocity in rectangular channels, and English manufacturers have provided machinery for clearing detritus from channels of approximately rectangular form.

The channel that is easiest to clean by either mechanical dredger or travelling grit-pump is not rectangular but one with sloping sides. Theoretically a channel of parabolic cross-section will have a constant velocity if controlled by a rectangular standing-wave flume, the invert of which is level with that of the parabolic channel. This has a distinct advantage because the flume can serve also for measuring the flow through the works. Channels of parabolic or approximate parabolic cross-section have been constructed, but it is possible to design a channel with a W-shaped section with sloping sides having no curved faces, which has very nearly the same cross-sectional area of flow as a parabolic channel, of the same depth and surface width, at most rates of flow.

The general approximate formula for the discharge of a standing-wave flume is

$$Q = 1\cdot706BH^{1\cdot5}, \qquad (46)$$

where Q = discharge in cubic metres per second,
B = width of the contraction of the standing-wave flume in metres,
H = depth in metres from invert of flume to upstream water-level

(see also Formula 41).

The depth of the standing-wave flume will depend on the available head and the width will depend on the flow. Having tentatively determined these dimensions, one can find the width at various depths of a channel of parabolic cross-section, in which the velocity will be 0·3048 metre per second, by the formula

$$X = 4\cdot92Q/H \qquad (47)$$

where X = width of channel at water-level in metres,
H = depth of flow in metres,
Q = rate of flow at that depth in cubic metres per second.

By combining Formulae 46 and 47 we have

$$X = 8 \cdot 4BH^{0 \cdot 5} \tag{48}$$

Having found the proportions of a single standing-wave flume and a single parabolic channel that will take the maximum rate of flow, it is necessary to divide the width of the flume and the widths of the channel at its various depths by the proposed number of channels, for it may be found that a single channel does not have sides sloping steeply enough to keep the channel clean and to make the detritus gravitate towards the centre. To the decided number of channels should be added one stand-by.

Each channel should have its individual standing-wave flume, isolating penstocks so that it can be laid off for maintenance and, if necessary, a washout valve for emptying. The length of the channel should be sufficient for a particle of the smallest size to fall to the bottom during the detention period. This means that to settle a particle of 0·2 millimetre diameter, the length would have to be about seventeen times the depth from top water level to invert. Preferably it should be longer to allow some storage of detritus in the invert: the channels at Crossness Sewage-treatment Works are, for example, about twenty-nine times as long as they are deep.

To transform the parabolic cross-section to W-shape, a parabola should be drawn on graph paper and a shape like a W, with the central ridge coming up to about half water depth, should be experimented with until a satisfactory form with all the sides at equal slopes has been found.

Detritus channels of actual parabolic cross-section can be cleansed by travelling dredger or by travelling grit pump, the suction of which swings from side to side to the radius that approximates to the invert curve of the channel: the water containing detritus is discharged into a grit-separating vessel carried on the travelling bridge, and from this the detritus can be dumped into wagons. Alternatively, the detritus may be discharged into a channel by the side of the detritus channel and gravitated to a lagoon, grit washer or grit separator. A W-shaped detritus channel requires a pair of grit pumps mounted on the same bridge. The bridge should move at not more than 75 millimetres per second upstream or 300 millimetres per second when going in the direction of the flow. Pumps should be capable of drawing off the detritus in a dilution of 98 % of sewage.

Detritus from such channels is fairly clean, having about 5 % or

6% organic content or about 3% after washing. The moisture content of detritus taken from a separator is about 30%. When dried the detritus weighs about 1360 kilogrammes per cubic metre. In one instance the quantity of detritus was found to be 10 kilogrammes per head of population per annum. Two-thirds of this quantity is the amount sometimes assumed.

Grease removal

Special arrangements for the removal of grease from sewage prior to the main treatment processes are seldom needed in Great Britain, for the sewage does not contain quantities of grease large enough to interfere with treatment except, perhaps, in some areas where some industry such as the manufacture of wool textiles discharges abnormally large quantities of grease to the sewers. A normal sewage does contain grease: for example, at works serving typical industrial and residential areas, grease balls of various sizes up to or exceeding 150 millimetres diameter are liable to collect in the channels feeding the primary sedimentation tanks.

Grease is most easily removed from sewage by a Dorr-Oliver process in which the sewage is aerated and then passed to a circular covered tank where a partial vacuum is applied, causing air and other gases to be given off. The bubbles carry the grease to the surface, whence it is removed mechanically. Another method is to pass the sewage through a channel having a detention period of about 3 minutes at the maximum rate of flow, and to aerate it with small bubbles by the diffused-air method. The grease is carried to the surface by the bubbles and can be drawn off at the lower end of the channel. This process can be combined with detritus removal in aerated detritus channels, as already described.

11
Separation and Treatment of Storm Water

At sewage works taking flows from separate soil sewers to which there are no illicit connexions from surface-water drains, the amount of wet-weather flow in excess of dry-weather flow should not be great and there should be no need to provide tanks specially for the purpose of dealing with storm water. At these works the whole of the wet-weather flow is given full treatment by passing through the primary sedimentation tanks, aeration units and final sedimentation or humus tanks, and all the pipework, channels, etc., must be made of ample proportions to pass the maximum capacities of the sewers, which will have been designed to take up to four or six times dry-weather flow (see Chapter 2).

Where sewage works treat flows from catchments, served in whole or part by combined or partially-separate sewers, the rate of rainfall run-off during storms of even moderate intensity will be too great to be delivered as it arrives at the treatment works for full treatment, and it becomes necessary to pass any flow in excess of that which the works can take to stand-by or storm tanks.

The Royal Commission on Sewage Disposal recommended in their Fifth Report [21], *inter alia*, that:

> 'As a general rule, special stand-by tanks (two or more) should be provided at the works, and kept empty for the purpose of receiving the excess of storm water which cannot be properly passed through the ordinary tanks. As regards the amount which may be properly passed through the ordinary tanks, our experience shows that in storm times the rate of flow through these tanks may usually be increased to about three times the normal dry weather rate without serious disadvantage.
>
> 'The overflow at the works should be made from these special

tanks, and should be arranged so that it will not come into operation until the tanks are full.

'Special filters which are only used in times of storm are not usually efficient, and should not be provided.

'Any extra quantity of sewage arriving at the works in storms, which has to be filtered, should be treated on the ordinary filters, which should be made sufficiently large for the purpose.

'In most cases it will probably suffice to provide stand-by tanks capable of holding one quarter of the daily dry weather flow, and it will not be necessary to provide for *filtering* more than three times the normal dry weather flow.

'Under the arrangements which we recommend no storm sewage arriving at the outfall works would be discharged without some settlement.'

Some engineers have ignored the last recommendation and discharged all flows in excess of six times dry-weather flow without any treatment. There is nothing to be gained by this malpractice which, as has already been mentioned, is not only contrary to the interests of public health but is illegal.

In other respects the recommendations of the Royal Commission have been adhered to more strictly than is desirable. The full treatment of flows up to three times dry-weather flow and not more has been the almost invariable rule for works taking flows from combined and partially-separate sewers, regardless of what has been learnt during the last half century. This practice could be modified with advantage, for any adequate percolating-filter scheme can accept much more than three times dry-weather flow in wet weather. Sedimentation tanks are made so much larger in Great Britain than the capacities found satisfactory in America that they can pass six times the average rate of flow and still produce a good settled effluent, and experiments have shown that percolating filters increase in efficiency with increase of rate of flow per unit of surface area until a loading of about 5 cubic metres per day per square metre of surface is reached. This means broadly that a percolating filter of average depth treating an average-strength sewage can take about seven times dry-weather flow without detriment. These figures suggest that all percolating-filter schemes should be so arranged that, at discretion of the operator, storm flows of up to six times dry-weather flow can be passed to the works for full treatment in lieu of the three times dry-weather flow recommended by the Royal Commission. Similar arrangements could be made at activated sludge works for, although they do not benefit from heavy loading to the same extent as do percolating filters, they are often capable of taking increased flows of diluted storm sewage without marked deterioration in the quality of the effluent.

The Royal Commission recommendation that tanks capable of holding one-quarter of the daily dry-weather flow will suffice in most instances was the crudest guess and, as will be shown, very wrong. When one comes to examine the problem, one finds that while small tanks will often suffice for storing the spill-over from many small storms, it would often be necessary to have tanks able to accommodate 2 days' dry-weather flow to prevent any spill-over whatsoever.

Before this matter is discussed, the function of storm tanks should be clarified. Storm tanks are intended to store storm water in excess of the maximum which the works will pass and then, after the storm has abated, to return it to the flow for full treatment. But should the storm flow exceed the rate of flow that can be given full treatment plus the storage capacity, the excess will pass over the outlet weirs of the storm tanks to the river or stream. Whether or not this overflow of settled sewage is objectionable must depend on local conditions. The tendency has been to ignore pollution from this source, but it can be appreciable as, again, will be shown.

The arithmetic of storm-water treatment is more or less on the following lines. A study of some very large catchment areas, occupied by nearly 12% of the population of Great Britain, showed that the impermeable area of a combined sewerage system averaged about 39 square metres per head of population. This, according to Bilham's formula, gives a run-off of 1·25 cubic metres per day on the wettest day of the year. If the dry-weather flow is the average of 0·25 cubic metre per head of population per day (see Chapter 2) the total flow arriving at the sewage works on the wettest day of the year will be six times dry-weather flow.*

If both storm flow and soil-sewage flow were to arrive at the works at a steady rate throughout the day and six times dry-weather flow were passed to full treatment there would be no theoretical need for the provision of storm tanks at the average works treating flow from a combined-sewerage system. But, as already described in Chapter

* That this is the same rate at which storm water used to be discharged without treatment is pure coincidence.

2, the intensity of short storms is greater than that of long storms and there is one storm during the year which calls for the maximum storage capacity. The capacity required to store the difference between the run-off due to this storm and the rate of outgo can be calculated by Formula 25, substituting unity for N, and the difference between maximum rate of flow to be given full treatment and dry-weather flow for P. If this is done it will be found that, for an impervious area of 39 square metres per head of population, a dry-weather flow of 0·25 cubic metre per head of population per day and a maximum rate of flow for full treatment of six times dry-weather flow, the required capacity of the storm tank to just prevent spill-over once a year will be a little over 2 days' dry-weather flow.

This is a large capacity but not uneconomic, bearing in mind that the cost of the tanks would be much less than the extra cost of drains and sewers in a separate sewerage system. Also there are many localities where the sewers are on the partially-separate system or where parts only of the catchment are on the separate system, and it would be possible to store the whole of the storm water in tanks of much smaller capacity.

It should also be borne in mind that a long detention period equal to 2 or 3-days' dry-weather flow does not mean that the sewage will be stored for a long time and become septic: the tanks will fill quickly until the rainfall run-off becomes less than the flow being passed to treatment, after which the tanks will empty rapidly and usually will be quite empty by the end of the day.

There is good reason for considering storm tanks of large capacity. It would not be rational for a river authority to ask for an extra-high quality of effluent from a sewage-treatment plant when the *un-sampled* discharge from the storm tanks was producing as much pollution in 1 wet day as was being caused by the treatment works as a whole in 10 or more days of dry weather, in terms of $B.O.D._5$ plus a heavy suspended-solids load. This is, in fact, approximately the average circumstance on the occasion of the maximum storm of the year at works having storm tanks with a detention period of 6 hours' dry-weather flow and where flows in excess of six times dry-weather flow are discharged without treatment.

Formula 25 provides a quick way of calculating the required capacity of tank: a slower, more laborious method is illustrated in Table 39. The latter, however, which is based on the same average data as the foregoing calculation, shows not only that a detention period of 48 hours is required to virtually do away with any spill-over but also gives the quantities of spill-over for various detention periods and for once-a-year storms of various durations.

It will be observed that the storm lasting about 4 hours requires the maximum storage. It will also be observed that the amount of spill-over varies very little for detention periods between 1 and 6 hours and by quite a moderate amount between 6 hours and 12 hours. It is, therefore, quite impossible to say that any detention period between 1 hour and 12 hours is the right period: all are unsatisfactory, for between 6 hours and 1 hour the amount of pollution of the river on the wettest day of the year is in the region of the amount which would be caused during a week of dry weather.

Design of storm tanks

The traditional storm tanks were long, rectangular and flat-botttomed, in every way similar to the rectangular quiescent sedimentation tanks that were common early in the century. At first they were manually cleansed but later travelling-bridge machines were installed on existing tanks or became normal for new rectangular tanks. The tanks were provided with weirs at the inlet ends, the weir for each tank being at a different level from the rest so that minor storms would not foul all the tanks. Floating arms protected by floating scumboards were provided at the outlet ends of the tanks for decanting the contents to the sewage-treatment works for full treatment after the storm was over. The outlet weirs of the tanks were also protected by scumboards which floated in grooves. Decanting, sludging (or starting sludging mechanisms) were manual operations.

There is no reason why modern storm tanks should be designed according to tradition or be manually controlled. A much more satisfactory arrangement is to construct circular tanks with simple rotating sludging mechanism. These should have inlet weirs at different levels (all below that of the outlet weirs) so that each will come into use in turn, and peripheral outlet weirs protected by floating-ring scumboards. At the end of the storm, back-flow should take place automatically as soon as the water level upstream of the tank becomes lower than that inside the tank, and the contents should be drawn down to a level above that of the settled sludge. At this point the sludging mechanism should automatically operate and clean out the tank.

SEPARATION AND TREATMENT OF STORM WATER 159

While storm tanks are usually located at the sewage-treatment works, they are sometimes provided elsewhere on overloaded existing sewerage systems in positions where, formerly, simple overflows would have been used. In such a case sludge would be returned to the sewers after the storm for re-separation at the works.

Storm-water separation

The relative level of the storm tanks and the method of storm-water separation depend on the relative levels of the incoming sewers and the sewage-treatment works. If the sewage has to be pumped to the sewage works, storm-water separation is most easily effected by having dry-weather flow pumps to deliver to the sewage-treatment works and, when these are beaten by the flow, storm-water pumps to deliver the remainder to the storm tanks. After the storm, as soon as the water level in the suction well of the sewage pumping station falls below the lowest cut-in level of the storm-water pumps, the contents of the storm tanks then may gravitate back to the suction well by automatic control.

If the drainage area gravitates to the sewage works, storm water is usually separated by module or orifice and storm-overflow weir. Various kinds of modules capable of working with sewage can be purchased or be specially designed for the purpose. These usually consist of a float-operated valve together with an orifice downstream, so arranged that when the head reaches a predetermined level above the orifice, the float causes the valve to close. The valve should be of such a kind that any tendency to choke causes it to reopen. The closing of the valve causes the sewage level upstream to rise until it spills over storm weirs and gravitates to the storm tanks.

With this arrangement it is necessary to empty the storm tanks by pumps of the right capacity to deliver the difference between the *maximum* dry-weather flow and the peak flow to be treated by the plant.

A simpler arrangement for use at small works is an orifice set at the invert level of the incoming sewer and downstream of storm weirs. This gives less accurate separation of storm water but may often prove satisfactory.

The discharge of orifices can be found by Formula 45 but with A as the area of the opening and H as the head in metres above the centre of the orifice if it has free discharge, or the difference of up-

TABLE 39. Spill-over from storm tanks

(Cubic metres for an impermeable area of 39 hectares and full treatment of six times a dry-weather flow of 2500 cubic metres per day)

Duration of storm (hours)	Rainfall run-off* (cubic metres)	Five times dry-weather flow† to treatment during storm (cubic metres)	Detention period (hours dry-weather flow)											
			Zero (No tanks)	1	2	3	4	5	6	9	12	18	24	48
1	4557	521	4036	3932	3828	3724	3619	3515	3411	3098	2786	2161	1536	—
2	5745	1042	4703	4599	4495	4391	4286	4182	4078	3765	3453	2828	2203	—
3	6538	1562	4976	4872	4768	4664	4559	4455	4351	4038	3726	3101	2476	—
4	7132	2083	5049	4945	4841	4731	4632	4528	4424	4111	3799	3174	2549	49
5	7628	2604	5024	4920	4816	4712	4607	4503	4399	4086	3774	3149	2524	24
6	8123	3125	4998	4894	4790	4686	4581	4477	4373	4060	3748	3123	2498	—
9	9213	4687	4526	4422	4318	4214	4109	4005	3901	3588	3276	2651	2026	—
12	10104	6250	3854	3750	3646	3542	3437	3333	3229	2916	2604	1979	1354	—
18	11491	9375	2116	2012	1908	1804	1699	1595	1491	1178	866	241	—	—
24	12482	12500	—	—	—	—	—	—	—	—	—	—	—	—

* From Bilham's figures in British Rainfall (1935).
† With six times dry-weather flow to treatment works and one dry-weather flow arriving at the works in addition to the storm flow, the flow into the storm tanks will be rainfall run-off less five times dry-weather flow, apart from variations of soil-sewage flow during the day which are neglected in the calculations.

SEPARATION AND TREATMENT OF STORM WATER

stream and downstream head if the orifice is completely submerged. The constant m is approximately as follows:

Sharp-edged plate	$m = 0.62$
Opening in wall at invert of channel	$m = 0.86$
Short length of pipe (diameter = half-length)	$m = 0.81$
Penstock	$m = 0.62$
Vena Contracta (size of orifice measured at upstream end of Vena Contracta)	$m = 0.97$

The last is constructed as follows:

Diameter at upstream end	$= D$
Diameter at a distance half D downstream	$= 0.78 D$
Diameter at a distance of $2D$ downstream from top end	$= 1.14 D$

The discharge over a simple plate storm weir can be calculated according to Formula 40.

Balancing flows from pumping stations

There is little to be gained by providing balancing tanks to even out peaks of flow at most sewage works. There have been occasions when some form of balancing capacity has been incorporated in the primary sedimentation tanks, but this added complication interferes with their performance and there is no evidence that it improves treatment.

There are, however, occasions when the flow from a pumping station to a very small treatment plant needs to be balanced to prevent the sedimentation tanks from being washed out by excessive turbulence and to ensure that the rate of flow to the percolating filters does not greatly exceed the optimum. This circumstance seldom occurs except where the treatment works are so small that they cannot accept the discharge of the smallest advisable size of unchokable sewage pump, which is about 0·38 cubic metre per minute. Then the best arrangement is to construct a balancing tank the capacity of which needs to be a little more than the working capacity of the suction well of the pumping station. The outflow from this tank should be controlled, to not more than the maximum rate of flow to the sewage-treatment plant, by some suitable module as already described or a constant draw-off floating arm. Where storm water has to be separated, the peripheral walls of the tank can serve as overflow weirs. The floors should have hopper bottoms with sides sloping at 60° to the horizontal to facilitate sludging and should be provided with sludging valves or penstocks. The tank can be used as a detritus tank and screen chamber.

12
Sedimentation

By letting sewage stand in a state of comparative quiescence suspended solids are enabled to fall to the bottom to form sludge or rise to the surface and collect as scum. This process, applied to average crude sewage, will effect upwards of 70% removal of suspended solids in well-designed municipal sedimentation tanks and, as the suspended solids have a considerable biochemical oxygen demand, a reduction of B.O.D.$_5$ value of about 42% is to be expected. As the capital and operational costs of sedimentation tanks do not amount to a great part of the outlay on sewage treatment, the process of settlement is very economical.

Sedimentation is essential in any percolating-filter scheme because percolating filters of normal proportions are liable to choke with sludge if not so protected. Settlement should also precede land treatment and is usual before activated-sludge treatment, although the necessity there is not so great.

Settlement of particles

When water is not very turbulent those solids which have a greater specific gravity than water subside, their motion being resisted by the viscosity and inertia of the water. Very small siliceous particles, of less than 0·13 millimetre diameter, fall according to Stokes's Law, the velocity varying as the square of the diameter. Large siliceous particles, of 10 millimetres diameter or more, are resisted by inertia and the velocity of settlement varies as the square root of the diameter. Particles with diameters between these values are noticeably affected in motion by both viscosity and inertia: in this range, siliceous particles fall at the speeds given in Table 36. There are also particles so small that they will not settle at all unless chemical precipitants are applied.

Sewage contains suspended solids of various sizes from the finest colloids to anything that can pass down the sewer. Screening and detritus settlement remove the largest of these, and so the sedimentation tanks have to remove mineral particles mostly smaller than settled in the detritus channels and organic materials both heavier and lighter than water but small enough to have passed the screens. As these various substances sink or rise at different rates, a greater proportion of solids will pass on with the tank effluent if sedimentation is applied for a short time only than would after long-period sedimentation.

The settlement of sewage is not as simple as that of detritus, for chemical and physical changes are taking place all the time. In particular there is flocculation of the finer particles which occurs naturally and can sometimes be successfully encouraged. This is the clinging together of small particles to form larger 'woolly' floccules which will sink more rapidly (see Table 40).

TABLE 40. *Effect of flocculation*

Fluid	Flocculating time (minutes)	Settling time (minutes)	Suspended solids (milligrammes per litre)
Raw sewage	—	—	368
Raw sewage	0	60	166
Raw sewage	30	30	103
Raw sewage and activated sludge	—	—	497
Raw sewage and activated sludge	0	60	132
Raw sewage and activated sludge	30	30	44

Flocculation can be encouraged in upward-flow tanks on the principle of the Dortmund tank. The floccules fall contrary to the upward velocity and sweep with them the smaller particles which otherwise would be carried away with the effluent. Flocculation has been encouraged by passing 'picket fences' of small-diameter rods through the liquid to push the particles together. The control of pH is necessary to produce the best results when some precipitants are used to assist flocculation.

Curves of settlement

The combined effects of the different falling rates of the various particles, flocculation, turbulence in the tank and all other factors, produce various rates of settlement. These can be expressed by curves relating degree of settlement to detention period for various initial densities of suspensions. Several such curves have been plotted and it has been found that, over the detention periods that are practicable at sewage works, there is no critical point which can decide the economic detention period. The author, having examined a number of curves, found that the following formula was reasonably representative of normal conditions of primary settlement:

$$S_2 = \frac{S_1}{C_1 \times \log S_1 \times D^n}, \qquad (49)$$

where S_2 = suspended solids in tank effluent in milligrammes per litre,
S_1 = suspended solids in crude sewage in milligrammes per litre,
C_1 = a constant, which in the main experimental data was approximately 1·1 and, in another case, 0·725,
C_2 = a constant, which in the main experimental data was approximately 10,
D = detention period in hours,
n = $(\log S_1)/C_2$ and in the second case mentioned, n was 0·4.

(Logarithms are to the base 10.)

Continuous-flow sedimentation

Early practice was to remove suspended solids by quiescent sedimentation for 2 or 3 hours. The sewage was run into a tank and allowed to stand, and then decanted with the aid of a floating arm leaving the sludge behind. This required several tanks, never fewer than four and preferably eight, for the processes of filling, standing and emptying, and usually necessitated manual control. Later it was found that a smaller capacity and much less trouble were made possible by continuous-flow sedimentation in which sewage was introduced into a tank and drawn off over weirs continuously, leaving the sludge behind.

It will be noted that detention period, or capacity in relation to flow, was the only factor considered by the Royal Commission and engineers in Great Britain. But long before the publication of the Fifth Report of the Royal Commission [21], which dealt chiefly with the relative merits of the various methods available for the purification of sewage, A. Hazen originated in America a theory of settlement which was to the effect that the shallower the tank, the shorter the distance the particles had to fall to become settled: therefore settlement depended mainly on the surface area of the tank. Experiments have shown broadly that surface area is more important than capacity, and it is now accepted that tanks having inadequate surface area will give inferior results. Nevertheless capacity is of importance in continuous-flow sedimentation, for turbulence must be reduced to a minimum where efficient settlement is wanted, and excessive turbulence is unavoidable in a very small tank.

It is virtually impossible to completely eliminate turbulence from a continuous-flow sedimentation tank and the degree of turbulence affects results. The velocities of turbulence must have upward and downward components and upward velocities must lift, from the lower parts of the tank, sewage containing more solids than above and, therefore, cause mixing of the contents of the tank, reducing sedimentation.

There are two ways in which this effect of turbulence can be combated. First, as much energy as practicable can be dissipated before the sewage enters the tank: second, the tank can contain a large mass of sewage to absorb, with the least disturbance, the kinetic energy that remains. To reduce the energy of velocity-head, the sewage can be passed from the inlet pipe to a chamber, baffle box or 'eddy bucket' from which it passes at a low velocity into the main body of the tank.

Large capacity is the simplest but most expensive way of obtaining efficiency. In Great Britain, where very large capacities were once usual, the tendency is still to use larger capacities than elsewhere. At the present time a detention period of 6 hours dry-weather flow for primary sedimentation tanks of works treating up to three times dry-weather flow is common practice, whereas in America detention periods of 1 to 3 hours have been recommended, 2½ hours being the most usual and 4 the maximum.

Calculations of economic detention periods made by the author suggested that American practice is more economically sound than the British. The economic detention period for primary sedimentation tanks preceding percolating-filter treatment was found to be 2½ hours dry-weather flow; any detention period exceeding 1½ hours dry-weather flow would be uneconomical for tanks followed by surface aeration,* and 1 hour was the maximum for a diffused-air scheme.

The choice of tank capacity is not purely one of settlement efficiency; it can be a matter of convenience or be determined by other factors. The economic size is by no means critical, there being very little difference in overall capital costs of schemes having primary tanks with capacities ranging from 1 to 10 hours detention period and in which the size of the secondary treatment works has been adjusted in accordance with the theoretical strength of the settled effluent. Thus it will often be found that the large tanks as used in Great Britain are not uneconomical, while they can be much more convenient to design and operate.

Inlet effects

The design of the inlet of a sedimentation tank is of great importance, for a bad inlet can ruin the performance of what would otherwise be a satisfactory tank. If sewage is introduced over a weir and allowed to fall as little a distance as 25 millimetres, the energy due to this fall will make a noticeable difference to the efficiency of the tank. In one instance a fall of 225 millimetres reduced sedimentation efficiency to below 50%; alterations to dissipate the energy outside the tank brought efficiency up to normal for the detention period of the tank.

A second effect of the inlet is that it directs the incoming flow of sewage, influencing the circulation of the sewage in the tank. Contrary to early opinion, sewage does not flow directly from inlet to outlet, but except in very rare circumstances, major eddies or rotations are set up in either horizontal or vertical planes. If, for example, sewage enters a tank by falling over a weir, it passes straight to the bottom, hits the floor, is diverted towards the outlet end, hits the end

* This was before the introduction of the Ames Crosta Mills high-intensity aerating cone.

wall and rises to the outlet weir. Here, part of the rising sewage passes over the weir and part returns along the surface back towards the inlet: thus a major vertical eddy is produced. Submergence of the inlet weir to direct the flow towards the outlet will cause an eddy to rotate in the opposite direction, that is, along the surface, down at the outlet end and then back along the floor towards the inlet. It is not possible to partially submerge the weir and cause a straight flow through the tank; if the weir is slowly altered from free falling to submerged, it is found that there is a critical point at which the contents of the tank change over from rotating in one direction to rotating in another.

Direction of rotation is influenced by difference between the specific gravity of entering sewage and that of the sewage in the tank. The flow entering a final sedimentation tank for activated sludge is almost certain to fall to the bottom under the weight of the sludge, and therefore the tank should be designed to facilitate, not oppose, this tendency. Sewage entering a primary sedimentation tank is less heavy with sludge and, if warmer than the contents of the tank, may flow over the surface.

Apart from the foregoing effects, if the sewage is brought into a tank by a number of inlets spaced across the inlet end, one or more jet eddies may be set up. In one instance where a tank was fed by penstocks from a channel that flowed across the end, a component of this cross-velocity was sufficient to visibly rotate the whole content of the tank in one great eddy.

The fact that the sewage usually flows down the tank and then returns to the inlet means that the velocity of flow through the tank must be greater than the 'theoretical' velocity as calculated by dividing the flow by the cross-sectional area. If this velocity were high enough it would, as does the velocity in a sewer, prevent or at least act against settlement. But it seldom happens that a sedimentation tank is made so long in proportion to cross-section that there is sufficient scour to need consideration.

Types of sedimentation tank

The main types of sedimentation tanks are as follows

1. Hopper-bottomed or conical-bottomed tanks with central inlets and peripheral weirs. These include the true Dortmund

upward-flow tank and the much more common design in which flow is outwards from the centre and slightly upwards.
2. Circular tanks with flat or slightly sloping bottoms or with bottoms sloping as steeply as 30° to the horizontal. These tanks are always sludged by some type of rotating mechanism.
3. Rectangular, longitudinal-flow tanks, at one time always hand-swept but now sludged by mechanical sweeping gear.

The type of tank should be chosen according to circumstance. At small or moderately small works, hopper-bottomed or conical-bottomed tanks have the advantages of being simple and not requiring mechanical raking gear. But such tanks are not to be recommended in cases where, owing to the size of the works, the number of individual hopper bottoms to be sludged would be too great for a man to deal with all the tanks in a day without the risk of drawing off large quantities of sewage in addition to sludge.

Sludge storage capacity

When sludge is withdrawn constantly, as in the final sedimentation tanks of an activated-sludge scheme, it needs to be pushed into a sump of small dimensions only from which it is delivered under hydrostatic head. The sludging of primary sedimentation tanks is different for, although sludging can be automatic and there are some means of arranging constant withdrawal, it is usual at works of all magnitudes to accumulate sludge in the tank until some specified time, when it can be discharged under observation. At small works primary tanks may be sludged once a week only, and provision should be made for a 4-day interval at least. At large works daily sludging is desirable; more frequent sludging has been practised at some works but can be detrimental, for if sludge is withdrawn twice a day one sludging must be at a time when thin, watery sludge only is available.

This manual sludging means that primary sedimentation tanks should have sludge hoppers capable of holding 1 to 4 or more days' sludge accumulation according to the size of the works and the design of the tanks; if this is overlooked the best results cannot be expected.

When sludge is withdrawn from hopper-bottomed tanks or the inverted-pyramid or inverted-cone sumps of flat-bottomed tanks, it does not simply come away as the valves are opened; sewage is drawn down through the centre of the mass of the sludge, and sludge is left hanging up on the sides of the cones or pyramids. This tends to produce a thin, watery sludge unsuitable for passing to sludge treatment without de-watering. But a thick sludge can be obtained by sludging carefully, drawing off a little sludge at a time and stopping when the sludge appears to be diluted by sewage. This process, known as 'bleeding off' the sludge, should be carefully practised at all sewage works where sludging is manual.

Design capacities and proportions

During dry weather both the strength of the sewage and the suspended-solids content vary with the rate of flow in a proportion dependent on local circumstances. For example, in one large town it was found that the peak rate of flow was about twice the average at midday and the strength twice the average shortly after that time, both reducing in almost direct proportion as the day went on. Because of this it can be said that usually about two-thirds of the sewage solids arrive at the works during 8 hours of the day. Accordingly, it is not unreasonable to design sedimentation tanks on the rate of flow during this period and not on the average daily rate of flow. Allowing for variation of flow from day to day also, a general rule can be suggested, namely that the surface area and cross-section of sedimentation tanks should be based on three times the average dry-weather flow, not just dry-weather flow. In storm times velocities may greatly increase if, as the author has suggested, works are to treat six times dry-weather flow. But then dilution is to be expected and therefore it is not suggested that, as a general rule, tanks taking such flows should be made of greater proportions than required at works taking a maximum rate of flow equal to three times dry-weather flow.

The following figures are suggested for tanks used according to English practice:

1. The inlet should be so designed that the velocity of the sewage at the point of entry does not exceed 70 millimetres per second at dry-weather flow and it should be directed as required to maintain the desired major eddy.
2. The surface area of all primary sedimentation tanks should be at least 1 square metre for every 10 cubic metres per day dry-

weather flow regardless of shape of tank or type of flow. It has been suggested that the surface area of humus tanks needs to be only one-half this figure and that that of final tanks for activated sludge works should be no more than three-quarters of this figure.
3. Longitudinal velocities or, in the case of tanks with central inlets, outward velocities, should not be excessive. Opinion varies considerably as to what are the reasonable limits but it is seldom difficult to obtain a low horizontal velocity in tanks of otherwise satisfactory proportions. The author's experience is that in unusually long rectangular tanks, very good results were obtained with a theoretical velocity of 4 millimetres per second at dry-weather flow.
4. Outlets should be in the form of suitably placed weirs as far as possible from the inlet and, where necessary, protected by scumboards which are not too near the weirs and not too deep.
5. There should be adequate capacity for storage of sludge without interfering with the flow through the tank.

Pyramidal-bottomed tanks

The simple, effective, pyramidal-bottomed tank is used at more works than any other type and almost exclusively at modern small works. Yet it can be so badly designed as to function inefficiently or even fail to settle any sludge at all.

The main consideration is the capacity for sludge storage. What can be overlooked is that, if the inlet of a truly pyramidal tank enters half-way down the tank, the capacity for sludge storage will be less than one-eighth of the total capacity of the tank, which is totally inadequate. A good procedure is to calculate the sludge storage on the basis of the expected quantity of sludge, and to make sure that the inlet will be well above the space that the sludge will occupy and is so arranged that the settled sludge will not be disturbed.

It is British practice for the sides of pyramidal tanks or sludge hoppers to slope at 60° to the horizontal, which is steep enough. The capacity of a 60° pyramid is given by the formula

$$C = 0.289 X^3, \qquad (50)$$

where C = capacity in cubic metres,
X = length of base of pyramid in metres.

A satisfactory rule is to assume that nearly 30% of the capacity of the tank, or the lower two-thirds of the depth, is occupied by sludge at a moisture content of $97\frac{1}{2}$% and that this sludge has accumulated over a period of 4 days. If tanks are likely to be sludged as infrequently as once a week, some extra sludge capacity may be required.

Pyramidal-bottomed tanks should have central inlets with baffle boxes to keep any high-velocity downward flow from interfering with the settled sludge. The outlet should be in the form of peripheral weirs protected by scumboards.

Sludging should be under hydrostatic head. A pipe should be brought from the bottom of the pyramid and carried in a straight line to above water level to serve as a rodding-eye.* The rodding-eye can be inside or outside the tank but must be in a convenient position, for it will frequently have to be used. A branch should be taken off this pipe, in as short a length as possible, to a sludge manhole where sludging can be controlled by a screw penstock or a sluice valve suitable for sludge. This outlet should be between 1 and 2 metres depth below top water level; inadequate depth will make sludging difficult to start and too great a depth will make slow bleeding-off difficult to control. The outlets of sludge pipes must be in a position where the sludge that is being drawn off can be clearly seen by daylight on the darkest day for, in the process of bleeding-off, the man responsible must be able to see that he is drawing off thick sludge and not sewage.

Sludge pipes are usually made of cast-iron and are almost invariably 150 millimetres in diameter, both for primary sedimentation tanks and humus tanks. At very large works, sludge pipes of 200 millimetres have been used but it is doubtful if this was necessary. Diameters of less than 100 millimetres are not recommended.

Circular mechanized tanks

The circular tank with mechanical sludging gear is being used more and more for primary and final sedimentation and at smaller works than formerly. For the settlement of primary sludge, the inlet is central with a suitable baffle box and the weirs are peripheral, resting on the outer walls and protected by scumboards, which are required

* One would hardly think it necessary to mention that the rodding-eye should be brought above top water level and yet, on a number of occasions, this has been overlooked with the obvious results.

for humus tanks also. The floor of a primary sedimentation tank or of a humus tank is flat, or moderately sloping in conical form towards the centre where there must be a conical sump of sufficient capacity to store the maximum amount of sludge liable to come down in 1 day or the average likely in 2 days, whichever be the greater. For this purpose a moisture content of not less than 95·5% should be assumed.

The sludge is swept to the sump by electrically-driven raking gear of which there are several designs. This should rotate at a peripheral speed of not more than 1·2 metres per minute: if speed of motion is too high, the efficiency of sedimentation suffers. The sweeping mechanism is arranged to cause the sludge to flow towards the central sump: it also incorporates a device to push the scum up a ramp that is close to the peripheral scumboard.

Circular mechanized tanks for final sedimentation of activated-sludge works may be flat-bottomed if conditions make deep excavation expensive. But a preferred arrangement is for the bottoms to be conical and sloping at 30° to the horizontal towards a small sump at the centre from which the sludge pipe connects. This pipe must be large enough to take the maximum rate of activated-sludge recirculation, allowing for some of the tanks being laid off for maintenance. Some means is required for the control of draw-off from each tank and one of the best arrangements is a telescopic weir which, by means of a gun-metal screw and handwheel, can be raised or lowered. It is always advisable to calculate the sizes of these weirs and not to accept a catalogue article which may not be able to take the flow.

If the slope is at 30°, the raking mechanism needs to be no more than a chain that is moved round the tank by a rotating ring that runs on rails or wheels at the periphery. The sludge gravitates towards the centre and the chain prevents any from being left behind to go septic.

If the floors are comparatively flat and are swept by blades on rotating arms or bridges, the blades should be arranged to trail along the floor, otherwise they will fail to keep it clean. A method that has often been used to 'ensure' that there shall be slight clearance only between blades and floor has been to fit the blades with temporary wooden screeds and, by running the mechanism, screed a layer of cement mortar true to shape. This practice overlooked the fact that when the tank was filled with water, the arms of the scraper became lighter and were deflected upwards, leaving a gap of as much as 40 centimetres at the periphery between the blades and the floor.

Because the flow in final sedimentation tanks for activated sludge is down at the centre, outwards across the floor and up at the periphery, an appreciable improvement in effluent can be effected not by placing the outlet weirs at the periphery but by supporting them on cantilevers so that they are in the form of ring channels at a distance in from the weirs of not less than 0·7 of the depth of the tank at that point [1].

No scumboards are required in final sedimentation tanks for activated sludge.

Rectangular mechanized tanks

A battery of rectangular tanks occupies less space than a battery of circular tanks of the same effective capacity, and the arrangements for feeding with influent and for collecting effluent are always much simpler. On the other hand, the machinery required for sludging circular tanks is lighter, simpler and more reliable.

Rectangular tanks are sludged towards the inlet end where a number of hopper bottoms of adequate storage capacity must be provided to collect the sludge. Each should have its own sludge pipe with visible outlet and valve or penstock control.

There are several ways of removing scum, most of which necessitate drawing off large quantities of sewage with the scum. A method which the author has used is to have a scum blade that pushes the scum towards the inlet end of the tank and over a partially submerged ramp to fall into the channel that takes the discharge of the sludge pipe.

The travelling bridge that carries the sludging and scum-removal blades should not move faster than 1·2 metres per minute when sludging, but it may move at any convenient faster speed, e.g. 3·67 metres per minute, when the blades are lifted and it is running back to the outlet end.

There are two main types of machine used for sludging rectangular tanks, the 'travelling bridge' and the 'flight collector'. The former can be an expensive, heavy machine and, as there is no merit in continuously sludging primary sedimentation tanks, whereas the bridge-type machine can sludge the largest tank in 2 hours or less, it is common practice to supply a sufficient number of machines to comfortably sludge all the tanks in a day (plus stand-by) together with a transfer carriage for each machine by which it can be moved

SEDIMENTATION 175

laterally from one tank to another. A suitable lateral speed is 6 metres per minute.

The flight collector consists of a series of sludging and scum-removal blades carried on chains that run over sprocket wheels fixed to the walls of the tank. Motion is continuous at about 1·2 metres per minute.

Sundry details

Scumboards have been constructed of timber, slate, steel and reinforced concrete. Creosoted softwood is perhaps most usual because it is inexpensive, can be easily replaced and is not fragile. If hardwood is used, oak is to be preferred but elm should not be used as it rots quickly when partly in and partly out of water. The usual proportions are 380 millimetres wide by 38 millimetres thick. The boards are submerged about 300 millimetres. Steel scumboards are most frequently used for circular tanks because they are easily made to shape. For fixing scumboards to the tank, galvanized angle-irons are formed to shape and secured by gun-metal rag bolts.

It is always an advantage for a tank to be capable of complete emptying by gravity or to a pumping station* but it can be unduly expensive to make such gravity-emptying arrangements for deep hopper-bottoms. Large tanks can be provided with washout pipes by which they can be emptied down to the level of the lowest part of the main floor† after which, if it is necessary, deep sumps may be emptied by portable pumps, provided that they are not too deep to be beyond the possibility of emptying by suction. These are points to be kept in mind when designing washout arrangements.

Open channels supplying influent to or taking effluent from sedimentation tanks should be carefully proportioned to ensure adequate self-cleansing gradients without causing too great a loss of head where falls are limited. The best cross-section of channel for general use is that rectangle which is twice as wide as its depth from water level to invert. A simple rule to remember when designing such channels is that the discharge of a channel of this cross-section is about one-half of the discharge of a circular pipe or culvert of the same diameter and material laid to the same gradient (see *Escritts' Tables of Metric Hydraulic Flow*).

* Every sewage works except the smallest has a general-drainage pumping station that can deal with tank-emptying.
† See Formula 33.

13
Sludge Disposal

Quantities of sludge

The total quantity of sludge solids from all parts of the works at British sewage works averages in the region of 0·079 kilogrammes per head of population per day. This, at an average moisture content of 95·52%, gives 1·76 litres of sludge per head of population per day. About 72·5% of the dry solids are volatile matter.

When sewage is treated by settlement only, not only are there no sludges produced from dissolved solids by aeration but a large proportion (usually less than 30%) of the suspended solids in the sewage is lost with the effluent. (Where there is secondary treatment these solids become more easily settled and are mostly retained.) In such works the quantity of dry solids averages 0·0545 kilogramme per head per day which, at an average moisture content of 92·5%, gives 0·726 litre of sludge per head per day.

The humus settled in the humus tanks of percolating-filter schemes consists of that portion of the solid matter which the primary sedimentation tanks failed to settle but which, by aeration, has become settlable, together with the debris of the flora and fauna of the percolating filters and any detritus of broken-down filter medium. The average quantity is 0·0268 kilogramme per head per day which, at an average moisture content of 99·23%, gives 3·48 litres per head per day. Of the dry solids 64·2% are volatile.

The average quantity of dry solids in the surplus activated sludge of a diffused-air plant is about 0·026 kilogramme per head per day which, at a moisture content of 99·33%, is 3·88 litres per head per day. Of the dry matter, 77·1% is volatile. It has been said that the quantity of surplus activated sludge from surface-aeration plants is twice as much as from diffused-air plants but, as about 80% of the surplus sludge is sewage solids that have not settled in the primary sedimentation tanks, as there appear to be some errors in the data so far available and as there is no good reason why there should be such a difference, this opinion should be viewed with caution.

The above figures may be used in design for estimating the quantities of sludge except where an analysis is available: then the analysis should be used. The total amount of sludge to be expected at the works is made up of the quantities of solids coming down the sewer, plus the estimated solids produced by the aeration processes, minus the solids lost with the effluent and the solids removed prior to settlement in the preliminary treatment works. In estimating quantities of sludge from analyses, the figures that relate to average daily flow, and not dry-weather flow, are used. It is easy to calculate the weight of solids that come down the sewers and deduct the weight of solids lost to the river, assuming an average suspended-solids content in the effluent. Deciding how much new suspended matter is produced by the biological process is far from easy and estimates based on the figures at various works differ considerably, probably because the required figure is the difference between two much larger figures and any errors made in sampling or in the laboratory are liable to be multiplied by about five. Averages of the figures for 5 years' operation of Crossness Sewage-treatment Works gave 0.125 kilogramme of suspended solids produced from dissolved solids per kilogramme of B.O.D.$_5$ removed by aeration: this is the lowest but probably the most accurate figure that the author has found.

Owing to the very large differences of moisture content of sludge, which cannot be predicted with accuracy, it is always advisable to assume the maximum moisture content likely for the type of sludge concerned, in all design calculations of pipe and channel sizes, pumping-plant capacities, etc.

Changes of quantities of sludge due to dewatering may be calculated by Santo Crimp's formula

$$W_2 = \left(\frac{100-P}{100-Q}\right) W_1, \qquad (51)$$

where W_1 = original weight of sludge,
W_2 = weight of dewatered sludge,
P = % moisture of sludge before dewatering,
Q = % moisture of dewatered sludge.

The sludge from primary sedimentation tanks contains fine, inorganic detritus (largely siliceous silt), various organic solids such as animal and vegetable fibre and broken-down particles, and considerable quantities of oil and grease of mineral, animal and vegetable origin. The animal and vegetable fats are digestible and are largely responsible for the gas which can be collected from sludge digestion tanks. The mineral oils are not digestible and their presence renders primary sludge an unsuitable material for regular use as manure. The average specific gravity of dry solids in primary sludge is about 1·31 (see Table 41).

TABLE 41. *Specific gravity and specific heat of sludge at 16°C*

%Total solids	Specific gravity	Specific heat
1	1·0031	0·993
2	1·0062	0·986
3	1·0093	0·978
4	1·0124	0·971
5	1·0155	0·964
6	1·0186	0·957
7	1·0217	0·950
8	1·0248	0·942
9	1·0279	0·935
10	1·0310	0·928

Methods of sludge disposal

Various methods of disposal are used, depending on local circumstances or, not infrequently, the preferences of the engineer or local authority. In a survey made by the Water Pollution Research Laboratory [24] it was found that, of 142 works, 106 had sludge drying beds, 13 used chemicals for sludge conditioning, 6 used rotary vacuum filters, one used filter presses and one used the Rotoplug sludge concentrator. It is not stated what other methods were used. These figures clearly indicate that drying on beds is the most popular method.

The approximate costs of the various methods are given in Table 42.

The simplest means of disposing of sludge at very small works is

TABLE 42. *Comparative costs of sludge-disposal methods*

Method of disposal	Cost as a percentage of cost of drying on beds
Disposal in permanent lagoons	25
Drying on earth plots	50
Pumping to farm land	50
Drying on sludge beds	100
Shipping to sea	100
Filter pressing	150
Heat-drying of press cake	300
Vacuum filtration and flash-drying	670

to run it into trenches which are refilled and redug as required. At larger works, land can be ploughed in the direction of the contours and sludge discharged onto it and allowed to dry, after which the ground can be harrowed and reploughed. At one large works this method has been adopted and the product sold as a substitute for top soil. The amount of land used was about 1 hectare per 12 000 head of population (see also Table 34).

In a few instances it has been possible to dispose of wet sludge by pipeline to nearby farmers. The difficulty is to discharge the sludge all the year round, for farmers require it at certain seasons only.

Sludge-drying beds

Properly constructed beds for drying sludge are rectangular, shallow beds with concrete floors, brick or concrete walls, underdrains and a filter medium. The underdrains can be constructed of circular or half-round agricultural tiles of 75 millimetres internal diameter laid to herring-bone pattern. Alternatively, a complete false floor of filter tiles of the type employed at waterworks may be used. To facilitate drainage the floor should be laid to a gradient of 1 in 200.

Over the agricultural drains or false floor is laid a bed of not less than 225 millimetres thickness of clinker or similar material, graded from 40 to 25 millimetres diameter, brought to a level surface (not sloping from the sludge inlet) and topped with a 40-millimetre thickness of fine clinker which will pass a 12·5-millimetre sieve but be held on a 6-millimetre sieve. This material has to be replaced from time to time because part of it is unavoidably lost when the dried sludge is removed.

The beds should be so constructed that a total of 250 to 300 millimetres depth of sludge can be discharged on to them and each bed should be so proportioned that it can be filled by one day's sludging; poor results are obtained if wet sludge is discharged on top of sludge that has partially dried. It follows that it is necessary to have a considerable number of small beds, not just a few large beds, to permit the regular sludging of tanks and the rotation of filling, drying and removal of dried sludge

Sludge is brought to the beds by open channel or pipeline controlled by valves or penstocks and laid to the minimum gradients given in Table 35. The minimum diameter for gravity sludge drains is usually limited to 150 millimetres, but larger diameters are not used unless justified by the flow.

At the inlet to each bed there should be an apron of concrete to protect the medium from scour. Each bed should have a weir penstock for the withdrawal of supernatant water. The drainage of this and the under-drainage of the beds should be discharged to a pumping station for delivery to the flow of sewage arriving at the works, for sludge liquor must be treated because of its strong biochemical oxygen demand.

After sludge has been discharged onto the beds it drains, dries at the surface and cracks, and can be lifted by manual labour or machine after a week or so according to the weather. At small works the sludge is manually dug out, using barrows and planks to protect the clinker. For large works there are various machines, in particular those made by Norstel & Templewood Hawksley Ltd., by which the sludge may be mechanically lifted and transported from the beds.

Over the years practice has changed and much larger areas are now used for sludge-drying beds than formerly. At one time it was usual to allow about 1·2 square metres for every ten head of population, but probably the most usual figure now is 2·4 square metres per ten persons.

Sludge drying can reduce the moisture content to about 55%. A cubic metre of dried sludge weighs about 930 kilogrammes.

Sludge-liquor pumping station

At some works a pumping station is installed to deliver humus back to the flow of sewage for settlement in the primary sedimentation tanks. At other works humus may be dried on sludge beds reserved for the purpose. Where there is a pump for delivering humus this may also serve to lift sludge liquor to the incoming flow of sewage, in which case it should be sized accordingly.

The sludge-liquor pump should not be so large that at small works it will cause disturbance in the primary sedimentation tanks, but it must be big enough at the smallest works to produce a self-cleansing velocity in a rising main of 100 millimetres diameter. The pump must have sufficient capacity to discharge the rainfall run-off of the sludge beds when they are empty (see Chapter 2), or deliver the maximum amount of sludge liquor likely to result from 1 day's sludging, it being assumed that the sludging is completed in 6 hours. If the station acts also as washout pumping station it should be capable of emptying one sedimentation tank or one humus tank in 6 hours.

The pump need not be capable of taking the peak rate of rainfall run-off provided that the suction well is of sufficient capacity to store the storm and that the pump is automatically started and ready to take rainfall run-off at any time. (For required capacity of suction well for storage of storm water see Formula 25.)

Vacuum filtration

A vacuum filter was a name given to a drum that rotated partly immersed in sludge and which internally was divided into sectors. A partial vacuum was applied through a rotary valve to those sectors which were under the sludge and emerging from it, then pressure was applied to loosen the cake that had formed during the vacuum period. Finally the cake was removed by a scraper set just clear of the filter fabric which usually consisted of metal gauze.

The disadvantage of filtering sludge is that the filter medium, no matter what it is made of or how coarse it is, rapidly becomes choked with fine particles, after which results may become very poor. If, for example, sludge is filtered through filter paper to produce a cake of the thickness obtainable on a vacuum filter, the cake tipped off and filtration repeated, the paper will become almost entirely choked after six such filtrations. But, should the paper then be turned over and reused, the backwash cleanses it and it may be used for another six times, turned over and reused, and so on indefinitely.

It is difficult to devise a practical mechanical filter that can be so cleansed by backwashing, but the difficulty has been overcome in the Komline–Sanderson 'Coilfilter' which is supplied by Dorr–Oliver Co., Ltd. In this filter, the fabric is not reversed but is nevertheless adequately washed. The medium consists of two layers of small-diameter stainless-steel helical coils. Each coil has its ends connected together so that it forms a continuous loop carried around the filter drum and the discharge and guide rolls of the mechanism. In the lower layer the coils lie side by side but not hard together, and in the upper layer they lie on the lower coils above the spaces between them.

During the cake-formation and cake-drying cycles, these coils are wrapped around the slowly revolving drum, the lower layer being in grooves in the transverse division strips of the drum. During the cake-discharge cycle, the coils leave the drum and are separated from each other by two discharge rolls in such a way that the cake is lifted off the lower layer and discharged from the upper layer of coils by means of a tine bar. Both layers are then washed by spray nozzles and reapplied to the drum by groove-aligning rolls: final effluent can be used for this washing. As the coils travel over the discharge rolls they bend first in one and then in the other direction, while being washed at the same time. This ensures that the filter medium is clean before the next dewatering cycle.

The filters are made in ten sizes from 5·2 square metres to 54 square metres nominal area. The nominal area is the diameter of the drum $\times$ the effective width $\times \pi$. The diameter and width of the smallest size quoted are 1·83 metres and 0·915 metre respectively and of the largest size 3·5 metres and 4·88 metres respectively.

The average output when handling a mixture of undigested primary and humus sludge is in the order of 14·6 kilogrammes of dry solids per square metre of filter area per hour. This figure will vary with the type and character of sludge being filtered, for example undigested primary sludge alone will give a higher average yield with less chemicals, and undigested primary sludge plus surplus activated sludge will give a lower yield with more chemicals.

For determining the size of a filter for works where overtime is not worked, it is usual to base the design on one 8-hour shift only per day for 5 days per week, the actual filter-working time per 8-hour day-shift being 7 hours to allow for starting up and shutting down.

It is usually desirable to pre-treat sludge with 5% of ferrous sulphate and 10% of lime relative to the dry matter in the sludge, or such other proportions as may be found effective with the particular sludge that has to be filtered. The filter cake has a moisture content of 80% or a little higher.

Filter pressing

Filter presses, which were widely used early in the century, consist of a number of cast-iron plates screwed together with coarse filter cloths between them. The spaces between the metal and the cloth form narrow cells into which on one side of the cloth the sludge is pumped under a pressure of about 4·2 kilogrammes per square centimetre (about 414 kilonewtons per square metre). The water escapes through the cloths and by draining grooves on the other side, and falls into a channel below. The plates are then unscrewed, the cake removed and the process repeated. The whole operation takes about three-quarters of an hour. The filter cake has a moisture content of about 75%. The cake can be further dried by stacking.

Shipping to sea

Where sewage works are on navigable waters sludge can be shipped to sea at reasonable cost, and this can well be the most satisfactory method of disposal. The sludge must be taken far from land to some place where floating solids will not drift back to the foreshore. It has been usual to dump the sludge into deep water, but the fact that the water is deep in the proximity of land masses can mean that there are high-velocity currents liable to move the solids and this may not be desirable. The presence of a mud bank, on the other hand, may suggest that they will remain where deposited.

For disposal at sea, it is necessary to have a jetty equipped with loading arms and special vessels. The sludge can be delivered by gravity from overhead tanks or direct from the sludge digestion tanks by gravity or pumping, according to the levels. The delivery arms are arranged to rise and fall and swivel to reach to the correct discharge position on all ships at all states of the tide and they have flexible pipes that can be dropped into the hoppers on the ships: design should be such that the flow of sludge does not cause these flexible pipes to lash out and distribute sludge over the deck. The sludge vessels should be provided with a number of sludge tanks, and valves by which the flow can be discharged to each tank in turn to trim the vessel. These tanks should be very well ventilated to prevent the accumulation of explosive gas. The outlets of the tanks should be designed to permit the sludge to gravitate into the sea leaving as little as practicable behind when large valves are opened.

In addition to the sludge tanks, the vessel should have adequate ballast tanks into which sea water can be pumped after the load has been discharged. There have been occasions when sea water has been pumped into the sludge tanks and then discharged from the ship on arrival at the jetty, but this causes considerable pollution of the river.

Sludge that is shipped to sea should be either undigested or well digested because partly digested sludge will gas too freely.

Heat treatment

The principal method of heat treatment of sludge is sludge digestion. This is so important that it is dealt with in a chapter on its own. Other methods include the Porteous process, in which disintegrated sludge is heated in pressure vessels then cooled, settled and dewatered by decanting the supernatant water, after which it is pressed to produce a cake of about 40% moisture content. This cake is sterile and free from weed seeds and therefore better as a fertilizer than untreated sludge.

The Zimmerman process is one in which the organic content of the sludge is destroyed by combustion in water at high temperature and pressure.

Sludge cake can be incinerated, using gas or oil fuel, in a multi-hearth incinerator, or in a suspension-type drier in which the sludge is mixed with dried sludge, passed through a flash drier and collected in a cyclone dust collector.

Sludge has been rendered more easily dewatered by slow freezing to produce ice crystals which push aside the solids. If, after remelting, the sludge is filtered without being unduly disturbed, very rapid dewatering by vacuum filtration is possible. The method, however, appears to be too costly for general use.

Sludge utilization

Sewage sludge as collected in the primary sedimentation tanks has manurial value but much less so than according to popular opinion. It has the disadvantages of containing weed seeds, harmful bacteria, far too much indigestible oil for continued use and, at some works, metallic salts liable to be injurious to crops. After digestion it contains a smaller quantity of oil, has less nitrogenous value and, being more fibrous, is a lighter material to apply to the soil. It also contains fewer pathogenic bacteria than raw sewage or sludge and many weed seeds are destroyed by digestion, with the marked exception of seeds of the tomato.

Humus from the humus tanks of percolating-filter schemes is a useful manure and may be separated and dried for this purpose in localities where there is a demand for it. Otherwise, as already mentioned, it is returned to the primary sedimentation tanks to be deposited with the primary sludge.

Activated sludge, collected separately, dewatered and heat-dried, makes a good manure for it has a high nitrogen content, contains little grease, is free from seeds and is believed to be free from harmful bacteria, but the cost of production of manure from it has usually been prohibitive. It should, however, be mentioned that there is a dewatering method which could well alter the economics of heat-drying, the Dissolved Air Flotation Thickening Process marketed in Great Britain by Ames Crosta Mills & Co., Ltd. It is claimed that this process will reduce the moisture content of activated sludge from 99·5% to 96%, and final moisture contents of between 95% and 93% have not been unusual. The method consists of dissolving air in water under pressure, adding this water (usually the recirculated effluent of the process) to the sludge and reducing the pressure. The fine bubbles given off on reduction of pressure carry the suspended solids to the surface where they are removed by a variable-speed mechanical skimmer. The material removed has a minimum solids content of 4%, contains a considerable amount of air and weighs about 700 kilogrammes per cubic metre. After withdrawal it is retained for 1 or 2 days in a tank which has a floor sloping at 60° to the horizontal or else is provided with scrapers; in this tank the air is given off.

The standard loading of the flotation tank is about 0·0325 cubic metre per minute per square metre of surface area and, as an average for design purposes, the solids load can be taken as 9·75 kilogrammes per square metre of surface area per hour, although the thickener will accept loading of two to two and a half times this figure. It is usual to allow for chemical-feed equipment to assist flocculation.

To all who propose to use sludge as manure, the following rules which were issued in California some years ago should be borne in mind:

'(1) Raw sewage containing human excrement shall not be used for irrigating growing crops.

'(2) Raw or undigested sludge shall not be used for fertilising growing vegetables, garden truck or low growing fruit or berries unless the sludge has been kiln dried, bed dried or aged in storage for at least 30 days, well digested (i.e. odourless, readily drainable and containing not over 50 per cent of the total solid matter in volatile form), or conditioned to the satisfaction of the State Board of Health.

'(3) Settled or undisinfected sewage effluents are banned for growing vegetables, garden truck, berries or low-growing fruit, or for watering vineyards or orchards where windfalls or fruit lie on the ground. Such liquid may be used for watering nursery stock, cotton, field crops such as hay, grain, rice, alfalfa, fodder corn, cow beets and fodder carrots, provided no milch cows are pastured on the land when moist with sewage, or have access to ditches carrying sewage. Use is also permitted for growing vegetables used exclusively for seed purposes.'

14

Sludge Digestion

When sludge is allowed to ferment on its own, at first acid digestion sets up, with the production of noxious gases, but eventually this is replaced by alkaline digestion which, once established, usually continues. In the intentional digestion of sludge, a quantity of actively digesting alkaline 'seeding' sludge obtained from other tanks or works is mixed with the raw sludge, and by this means the acid digestion phase is avoided.

When sludge is so seeded there is, at first, a period of rapid digestion with a high rate of gas production and an increase in pH value or alkalinity. Thereafter the rate of digestion decreases with the reduction of rate of gas production and of the rate of increase of pH value. Finally the production of methane becomes active and, if permitted, will continue until the digestible matter is exhausted.

The purpose of sludge digestion is primarily to reduce the quantity of sludge that has to be removed. Other advantages are removal of many pathogenic bacteria, the destruction of weed seeds and removal, to a large extent, of the oil content: 3 weeks mesophilic digestion will reduce the oil in the sludge from about one-quarter of the total volatile solids in the raw sludge to about one-fourteenth of the original volatile solids. Also digested sludge has an inoffensive odour sometimes described as 'tarry' and a light, fibrous nature which makes it much more suitable for application to the land than raw sludge. Finally, the gas produced by digestion has a high calorific value and is usually in sufficient quantity to provide the power required at the works for aeration, pumping and all other purposes, while the heat salved from the cooling water and heat exchangers of the engines is usually ample to raise the temperature of sludge to the optimum for digestion purposes.

Types of sludge digestion
There are three types of sludge digestion:

1. Cold digestion at the normal day temperature of a temperate climate or slightly above, for sewage is usually warmer than day temperature.
2. Mesophilic digestion in which the temperature is raised artificially and maintained, as near as practicable, between the limits of 32°C and 34°C.
3. Thermophilic digestion in which the sludge is maintained at a temperature between 44°C and 49°C.

Cold digestion

At 7°C (which is the minimum winter temperature of British sewage) digestion is inhibited but at 16°C (the summer temperature of sewage) digestion becomes appreciable. Thus it is possible in Great Britain to digest sludge without applying heat and the method is very practicable in semi-tropical countries. In Great Britain cold digestion is used at comparatively small works only, the average population served by any one plant being about 16 500, whereas the average population served by each mesophilic digestion plant is eight times this figure.

Cold digestion cannot be considered an outmoded method for there are very many installations in operation. On the other hand, these installations are not all as satisfactory as they might be, for the simple reason that, on the average, they have capacities too small to effect as much digestion as would be desirable. For satisfactory results a detention period of 3 months is required, which is 70% more than the present average capacity and a detention of 4 months would be more satisfactory. If it were appreciated that these larger capacities were necessary, the popularity of cold digestion would probably decline.

The shape of cold digestion tanks does not appear to affect results and any shape that is economical to construct can be adopted. It is advantageous to have digestion in two stages so that the comparatively rapid gas emission in the primary stage does not interfere with the formation of water strata in the secondary stage. The primary tanks should have a minimum capacity of 0·1 cubic metre to 0·134 cubic metre per head of population and the secondary tank a capacity of 0·064 cubic metre to 0·085 cubic metre per head of population. Provision should be made for the withdrawal of supernatant water and sub-surface strata of water at various levels in the secondary digestion tank.

Mesophilic digestion

Although thermophilic digestion is more rapid than mesophilic digestion, the former method is liable to be unstable and requires much more heat input; it is not used in Great Britain, where digestion temperatures exceeding 36°C are very rare. For this reason thermophilic digestion will not be discussed further. Mesophilic digestion is the most usual process at all except small works, and gas collection from primary mesophilic digestion tanks can be economic at large works.

Two-stage digestion is usual, gas being collected from the primary tanks only and stratified liquor from the secondary tanks. Probably the best way of determining the required capacity of primary digestion tanks is to allow 1 cubic metre of tank for every 1·55 kilogrammes per day of volatile solids in the sludge or, on the average, 1 cubic metre for every 2·14 kilogrammes per day of total sludge solids. This, at the average of 4·48% solids content in the sludge, works out at 21 days' detention or, on the average, 0·037 cubic metre of tank capacity per head of population. This capacity can be increased by 4/3 with advantage.

If there is any reason to doubt that a sludge with a solids content of 4·5% or more will be obtained from the sedimentation tanks, dewatering tanks should be provided. There should be two or more hopper-bottomed tanks each having a capacity of 1 day's production of sludge at the initial solids content and each provided with a floating arm for decanting top water. After the supernatant water has been withdrawn the dewatered sludge is delivered to the primary digestion tanks.

The average capacity of secondary digestion tanks following heated primary tanks, and intended to be capable of securing in the region of 30% reduction of sludge by careful decanting of sludge-liquor strata, is not less than 0·064 cubic metre per head, or about 0·09 cubic metre per head if a very good standard of dewatering is desired.

Solids reduction and gas production

With the above capacities and a temperature of 32°C in the primary tanks (but no heating in the secondary tanks) about 40·8% digestion of volatile solids (or on the average 29·6% of total solids) should be effected. About 0·96 cubic metre of gas per kilogramme of solids destroyed or about 0·0222 cubic metre of gas per head of population per day can be expected. These are all approximate average figures and very considerable local variations are to be expected. In particular, gas production varies not only from one installation to another but, at any one plant, may be as much as 10% above or below average for as long as a fortnight at a time and, while in design calculations it may be a fair assumption that an average of 0·0222 cubic metre of gas per head per day will be produced, the sizes of gas mains etc., and the surface area of primary digestion tanks, could well be designed on twice this figure.

The surface area of the digestion tank is important, for foaming can occur when the rate of gas emission exceeds about 10 cubic metres per day per square metre of tank surface and, to be safe, it is usual to restrict emission to one-half this figure. This limits the depth to which the tank should be constructed to about 6 metres, more or less, according to maximum estimated gas emission.

Gas produced by sludge digestion consists mainly of methane and carbon dioxide. There are also small quantities of nitrogen, e.g. $2\frac{1}{2}\%$, smaller quantities of oxygen, e.g. $\frac{1}{2}\%$ and traces of other gases. For design purposes it is usual to assume that the methane content amounts to 67% and, as the net calorific value of methane is 33 347 joules per litre, the net calorific value of an average sludge gas is 5336 kilocalories or 22 343 kilojoules per cubic metre. In practical terms this means that 1 cubic metre of sludge gas will, on the average, produce 1·86 kilowatt hours, assuming dual-fuel engines and generators of average overall efficiency.

The proportion of methane to carbon dioxide varies with the detention period and, if the detention period is reduced too much, the carbon dioxide content can become so high that the gas is rendered unsuitable for power purposes.

The waste heat that can be recovered in the power house and used for heating the sludge is appreciable. With the engines running at three-quarter load (which is usually allowed in calculations), about 19% of the calorific value of the gas used can be recovered from the cooling water and about 14% from calorifiers on the exhaust, giving a total of 33% of the calorific value of the gas.

With the production of gas there is a reduction of animal and vegetable fats and other organic substances, but this does not appear to be in direct proportion to the weight of gas produced. In many instances the recorded weight of gas is greater than the weight of organic matter destroyed and this has been attributed to a chemical reaction in which water has been decomposed and combined with oils to form methane and carbon dioxide. Other gases which are sometimes in appreciable quantity are nitrogen and oxygen: their presence may be due to accidental entrainment of air and denitrification on activated sludge coming in contact with other organic material.*

When estimating for power purposes, it is best to assume that the gas production is equal to the weight of solids digested, i.e. about 0·96 cubic metre per kilogramme of solids digested, but in designing gasholders, gas mains, excess gas burners, etc., it is safer to assume an average of 1·33 cubic metres per kilogramme, and a fluctuation of 20% above average.

Heat requirements of primary digestion tanks

In Great Britain it is necessary to raise the temperature of the sludge from about 7°C to not less than 27°C in the winter, or an increase of 20°C. In the summer it will not be necessary to raise the temperature more than from 15·5°C to 35°C at the most or an increase of 19·5°C. It should therefore suffice to raise the temperature of the sludge by 20°C at all times of the year, but it is usual to allow for a rise of 25°C. This means an input of 25 000 kilocalories (104 670 kilojoules) per cubic metre of sludge put into the tank.

In addition to the heat required for heating the sludge from sewage temperature to digestion temperature, heat input is necessary to make up loss by radiation. This can be estimated in two ways: first, by comparison with losses at similar works in a similar climate, and second, by heat-exchange calculations. There is some heat produced by the digesting process and this, as in the fermentation of wine or of manure on a dunghill, could be appreciable but as there are no

* At one site where there was no secondary treatment, a large-scale mesophilic and thermophilic experiment was made with temperatures varying from 25·6°C to 49°C and detention periods from 14·6 to 32·7 days; during the whole of these experiments the quantity of gas varied between the limits of 0·87 and 0·98 cubic metre per kilogramme of volatile matter digested and the average was almost exactly the same weight of gas produced as of volatile matter destroyed. Later, when activated-sludge and permanent sludge digestion plants were constructed, the recorded quantity of gas varied, over a period of years, between the limits of 1·16 and 1·33 cubic metres per kilogramme of volatile matter digested. It was known that a fault in the heating arrangements was causing entrainment of air and this must have been contributing to the extra quantity of gas, but no analyses of the gas have been published.

accurate figures available this is neglected. In a large-scale experiment made at the Beckton Sewage-treatment Works in which quantities of sludge input and input and output temperatures were regularly recorded, the amount of heat from sources other than the deliberate input via the heat exchanger was slightly greater, according to the instrument readings, than the radiation losses. In most other instances radiation losses have been found experimentally by stopping the sludge input and the heating and then recording the daily fall of temperature of the tank for a period. This method should be expected to give an experimental loss somewhat less than the loss due to radiation because it would include for some, but not for all, of the heat due to digestion, since the rapid digestion that takes place on the introduction of fresh sludge to the tank would have been eliminated.

Figures for heat losses by radiation from digestion tanks in Great Britain have been given as the equivalent of a drop of temperature in the whole volume of the tank content varying between 0·28°C and 0·83°C per day. These are extreme figures, and for design purposes it is usually sufficient to assume that the heat loss for digestion tanks surrounded by embankments and with concrete roofs is the equivalent of a temperature drop of 0·4°C per day, while the loss from concrete tanks with bare walls standing above ground level and steel gasholder-roofs without heat insulation can be taken as averaging the equivalent of a temperature drop of 0·65°C per day. With heat insulation, losses from exposed walls or gasholders can be very greatly reduced.

Theoretical heat-exchange calculations may be necessary when it has to be decided whether or not heat insulation is required or to determine the effect of such insulation. It should be mentioned that such calculations must not be taken as necessarily accurate, for there are too many assumptions involved and the amount of heat loss may eventually be found to be as much as 70% more or considerably less than as found by the calculations.

Radiation losses can be calculated by the formula*

---

* At the time of writing, British Standard recommendations are that heat should be measured in kilojoules, heat flow in kilowatts and thermal resistance in metre degree Celsius per watt. These are unnecessary complications in the case of sludge-digestion heat-requirement calculations, which involve neither mechanical work nor electric power, and therefore the method used here is not the same as the current British Standard recommendation.

SLUDGE DIGESTION

$$Q = A\left[\frac{t_1 - t_0}{(1/f_1) + (b/K) + (1/f_0)}\right], \qquad (52)$$

where Q = kilocalories per hour,†
A = area of wall surface in square metres,
t_1 = temperature in degrees centigrade on the hot side,
t_0 = temperature in degrees centigrade on the cold side,
f_1 = surface coefficient inside,
f_0 = surface coefficient outside,
b = wall thickness in metres,
K = thermal conductivity.

Suitable values for $1/f$ are given in Table 43 and for K in Table 44.

TABLE 43. *Reciprocals of surface coefficients*

Surfaces of contact	Reciprocal of surface coefficient* $1/f$
Sludge to methane	0·123
Methane to underside of gasholder	0·123
Gasholder to outside air, average weather conditions	0·05125
Sludge to inside of walls	negligible
Vertical wall to outside air	0·082

* These are 0·205 times the values used in former calculations to Imperial units.

The loss of heat is in direct proportion to the area of roof, wall or floor exposed and to the difference of temperature between the hot side and the cold side. It is in inverse proportion to the sum of the heat-resisting values of the materials between the inside and outside of the tank. Thus, if one is required to calculate the loss of heat from the sludge through the gas below the gasholder and from the gasholder to the outside air, the formula becomes

$$Q = \frac{A(t_1 - t_0)}{(1/f) + (1/f_1) + (1/f_0)}. \qquad (53)$$

In this instance the chief resistance to transfer of heat is in the surface

† 859·845 kilocalories per hour = 1 kilowatt.

TABLE 44. *Thermal conductivities of materials*

Material	K (approx.)*
Expanded polystyrene, fibreglass	0·03
Granulated cork, slag wool	0·04
Fibre-board, balsa wood	0·05
Loosely packed asbestos, plasterboard, oak, deal	0·14
Asbestos-cement sheet	0·25
Concrete	0·83
Heavy damp clay	1·03
Engineering brickwork	1·12
Loam	1·13

* These are 0·124 times the values used in former calculations to Imperial units, i.e. B.th.u, sq. ft. area, °F and inches thickness.

films of contact between sludge to gas, gas to metal gasholder and gasholder to outside air. Convection in the gas reduces heat resistance of the gas-occupied space and the heat resistance of the metal gasholder is negligible. By reference to Table 43 it will be seen that the value of $1/f$ for sludge to methane is 0·123, as is the value of $1/f_1$ for methane to underside of gasholder, while the value of $1/f_0$ for resistance on the outside of the gasholder surface is lower as a result of wind and weather, being 0·05125.

For the vertical walls with sludge on the inside the formula becomes

$$Q = \frac{A(t_1 - t_0)}{(b_1/K_1) + (b_2/K_2) + (1/f_0)}. \tag{54}$$

In this formula b_1 is the thickness of the concrete wall and K_1 the thermal conductivity coefficient for concrete, b_2 the thickness of any heat-insulating material and K_2 the thermal conductivity coefficient of the material. The value of $1/f_0$ for the resistance to loss of heat from the vertical surface of the wall is about 0·082.

It will be observed that additional values, b_3/K_3, b_4/K_4 and $1/f_2$, $1/f_3$ etc., can be added according to the various heat-resisting materials and cavities in the structure of roof, wall or floor.

The loss of heat through the floor is very speculative, for when the tank is first put into use there will be some considerable loss while the ground is being heated and, if there is movement of ground water, this loss will continue. If there is not movement of ground water, or particularly if the ground is dry, the great thickness of ground between the underside of the floor and the open air can be an effective barrier to heat transfer, but its value is very hard to estimate. In such a case the thickness of ground could, perhaps, be taken as equal to one-third of the radius of the tank plus the distance of the outer edge of the floor below ground level.

Methods of heating the sludge

Methods of heating the sludge to mesophilic temperature include hot-water pipes fixed to the insides of the walls, hot-water pipes carried on a slowly rotating stirring-mechanism, withdrawing the sludge from the tank and passing it through a heat exchanger, and passing the sludge through a heat exchanger inside the tank.

When the sludge in the primary tank is heated by means of fixed or slowly moving hot-water pipes the amount of heat exchange is about 49 kilocalories (about 205 kilojoules) per hour per square metre of external surface of heating pipe per degree centigrade difference of temperature between the hot circulating water and the heated sludge. The usual temperature of heating-water is in the region of 55°C. In the course of time, boiler crust forms on the *outside* of the heating pipes and reduces the rate of heat exchange, particularly if higher temperatures are used.

The rate of heat exchange is much greater when sludge is heated in some type of external heat exchanger through which the sludge passes at a velocity of about 1 metre per second, for the motion reduces film resistance. Heat exchange in the order of 244 to 294 kilocalories (about 1020 to 1230 kilojoules) per hour per square metre of heat-exchanger surface in contact with the sludge per degree centigrade difference of temperature between the sludge and the circulating water may then be expected.

In this category come the 'Simplex' sludge heater which is installed in a house outside the tank and the Dorr–Oliver 'B type' heater which is fixed in chambers on the outside walls of the tank. The 'Simplex' heater comprises a cylindrical body mounted horizontally and enclosing a helical tubular coil. The body is filled with hot water from the cooling water system and heat exchangers on the exhausts of the dual-fuel engines that are run on sludge gas. Sludge is withdrawn from the primary digestion tank at a high level, pumped through the helical coil, where its temperature is raised the required amount, and returned to the digestion tank at a low level. In the event of there not being sufficient heat available from the engines the water is heated by sludge-gas burners which are mounted on the front plate of the heater. The hot gases pass through fire tubes which extend longitudinally through the heater body and discharge to a chimney.

Ames Crosta Mills & Co., Ltd. supply machinery for both floating-roof and fixed concrete-roof tanks. The floating roof carries an electric axial-flow pump of special design and an uptake tube so arranged that the contents of the digestion tank can be turned over by sucking from the bottom and discharging at the surface. The pump is arranged to run at intervals under the control of a time switch. It can be reversed to suck from the surface and discharge at the bottom of the tank.

The floating roof permits the level in the tank to be varied considerably and, in design, it is usual to allow for the storage of 7 days' sludge production in that part of the tank through which the top sludge level varies. Gas is withdrawn from the floating roof and delivered to a separate gasholder. In the Simplex tank with a fixed concrete roof, heating of the sludge is effected by a water jacket surrounding the uptake tube in the centre of the tank.

The Dorr–Oliver 'B type' heater consists of a tube through which hot water is passed and a central core. The sludge is withdrawn from the tank at two different levels and also via a scum-removal horn, and pumped through wide annular spaces between the central core and the tube and between the outer face of the tube and the circular chamber in which it is housed, thereby taking heat from the hot water. The sludge is then returned to the tank at an angle calculated to effect mixing and cause the contents of the tank to rotate. Raw sludge from the primary sedimentation tanks is introduced downstream of the heater and mixed with the heated sludge before it is returned to the digestion tank. The heater bodies are so arranged that they can be lifted out by a special crane for the removal of boiler crust which can form on the outer surfaces. This is best removed by dipping the heaters into an acid bath and then into a neutralizing alkaline bath.

The Dorr–Oliver 'B type' heater can be used with tanks having fixed concrete roofs or with tanks having gasholder roofs; in the latter case the heater is known as the 'BGH heater'.

In the 'Heatamix' system sludge is recirculated inside the tank with the aid of gas-operated 'air-lifts' which are fitted with heaters and operated by sludge gas in lieu of air. In the installation at Mogden four 'Heatamix' steel tubes about 8 metres long and 300 millimetres diameter are fitted in a tank of 21·3 metres diameter and 4000 cubic metres capacity. The detention period during a test was 14 days.

The delivery of each 'Heatamix' is 6 cubic metres of sludge per minute for which 1·13 cubic metres per minute of gas are required per tube. For the whole installation 0·16 cubic metres per minute of hot water are required, raising the temperature of the sludge by 15°C. This gives a temperature drop in the water from 60°C to 40°C.

The bottoms of the tubes have bellmouths: the top ends turn over at right-angles 1·2 metres *below* surface level. It was found in the tests that to give effective disturbance of the scum, the direction in which these bends pointed was critical. The tubes are located 3·65 metres from the circumference of the tank. The bends of two tubes opposite each other discharge at right angles to the radii to spin the contents of the tank. One of the other two points at the tube ahead of it and the other to a point half-way between centre of tank and the circumference on the radius that passes through the tube ahead of it.

Shape of mesophilic tanks

A primary heated digestion-tank is usually circular on plan, for this shape makes possible economical construction in reinforced concrete or steel, is very suitable for use with gasholder roofs or the floating roofs of Ames Crosta Mills 'Simplex' type and is necessary for a tank with any form of rotating scraper or stirring mechanism. There is some difference of practice in the design of floors for primary digestion tanks. If the tank is fitted with a scraper mechanism arranged to draw the sludge towards a central outlet, as in some types of Dorr–Oliver tank, the floor will be flat. This not only simplifies construction but makes it easy for men to work in the tank without the danger of slipping whenever maintenance is required. In several designs where sludge is circulated by pump, as in the Dorr–Oliver 'B type' and 'BGH type' tanks or some of the Ames Crosta Mills 'Simplex' tanks, the floor has been sloped to the centre at an angle of at least 40° to the horizontal. This arrangement has, however, proved disadvantageous, for maintenance men cannot stand on the sloping floor; it is now thought better for the floor to be made to slope moderately at not more than 1 in 6 to the central outlet, and for the tank capacity to be calculated on the assumption that grit will form banks sloping at 40° to the horizontal towards the outlet, thus reducing the effective capacity.

If a rotating scraper mechanism is used, a scum-removal device is incorporated. This usually consists of a blade that pushes scum over a ramp near the periphery of the tank. Arrangements for scum removal are desirable in all primary digestion tanks, otherwise scum of an indestructible nature such as rubber held up by gas content, or corks, will accumulate over the years until finally the tank has to be emptied and a difficult scum-removal operation put in hand.

The shape of the secondary digestion tanks can be generally similar to that of the primary tanks and designed mainly from the point of view of economy of cost. A roof is neither necessary nor desirable. The outlet should be from the lowest point to permit discharge of the heaviest material and the inlet or inlets as far away as possible from the outlet. If the tank is circular with the outlet from the bottom of a conical floor, a number of peripheral inlets can be provided sufficiently submerged so that there will be no splash-down at any time liable to cause disturbance and aeration of the sludge. The incoming sludge, being warmer than that in the tank, should initially rise to the surface. The tank should be provided with a number of decanting valves at various levels and in various positions for removal of water bands.

Gasholder roofs

A gasholder roof on a primary digestion tank has one advantage over a separate gasholder: the alkaline digesting sludge protects the steel of the gasholder against rusting, whereas separate gasholders with gas seals of clean water do suffer from rust. Special rectangular gasholders have been made for rectangular digestion tanks, but by far the most satisfactory gasholder roof is the single-lift spiral-guided gasholder.

This needs to be of special design for the purpose. The internal trusses are arranged differently from those of a gasholder for town gas in that they must never at any time dip into the sludge for, should they do so, they will collect fibrous matter which would make them heavy and cause an excess of gas pressure.

The gasholders should be provided with several bolted accesses that can be removed at any time the tank is laid off for maintenance, a glass window with condensation wiper operated from the outside, a manhole trapped to prevent escape of gas and an anti-vacuum valve as a precaution against the gasholder being turned inside out by negative pressure should, by some accident, sludge be withdrawn at a time when the gasholder is resting in its lowest position; this catastrophe has occurred at several works but can be avoided by proper design and precautions in operation.

Opinions differ as to the required capacity of gasholder. Babbitt [2] suggests about 27 hours production for works where the gas demand is constant and all gas is used. Imhoff and Fair [12] suggest that the gas stored in the holder should equal the volume of fresh sludge added to the tank daily; this, for average English conditions, would mean a capacity equal to 30 hours production.

English practice has been to install much smaller gasholder capacities, e.g. from 4 to 8 hours gas production is not uncommon. In the design of the gasholders for the Beckton Sewage-treatment Works, the author allowed a capacity equal to 8 hours average gas collection, and at the Crossness Sewage-treatment Works about 7 hours gas collection, the actual capacity depending on convenience of design, a gasholder roof being provided on every tank. In both cases capacity was adequate.

Gas collection and utilization

Gas should be collected from each digestion tank by a pipe carried to not less than 1 metre above the highest sludge level to be clear of foam. The connexion from every sludge digestion tank and to every point of utilization of sludge gas or excess-gas burner should be isolated from the sludge pipework by a suitable flame trap of crimped stainless steel ribbon. Traps should be provided on all low points on the pipework to collect any condensation water. A connexion should be provided from a high-pressure water supply to clear the pipework should it become choked by foam.

It is usual to design gasholders and sludge-gas pipework on the assumption of a gas pressure varying between the limits of about 100 and 200 millimetres of water.

Excess-gas burner

The rate of consumption of gas for power purposes etc., must vary from the rate of production, for neither production nor consumption is steady. Large storage is uneconomic or impracticable and it follows that, should the power demand exceed the gas supply, either the demand must be reduced or power obtained from other sources such as the consumption of diesel oil in dual-fuel engines, or by buying power from the Central Electricity Generating Board.

When the supply of gas exceeds the demand, gas is passed to waste and, while this has been done by discharge to the air, it is usual to burn it in a specially designed burner that is put into operation manually on instruments showing that the gasholders are full, or nearly full or, in some instances, automatically.

There are several designs of excess-gas burner but the simplest, and generally most satisfactory, is a large bunsen burner. These bunsens are designed empirically. The size of the gas nozzle at the bottom, which can be determined from Table 45, must, at the largest works,

TABLE 45. *Discharge of gas through orifices*

Pressure of gas in millimetres of water	Cubic metres per minute per 100 square millimetres of orifice
100	0·186
125	0·208
150	0·228
175	0·246
200	0·263
225	0·281

be sufficient to take at least 20% more than the maximum estimated gas production for, when the works are first put into operation, it may be necessary to burn to waste the whole of the gas produced. At small works the burner should be able to pass twice the maximum gas production. The vertical tube of the bunsen burner should be from three to four times the diameter of the nozzle. There should be means of varying the diameter of the nozzle according to experience of operation.*

* According to mechanical engineering textbooks, a burner as above described will not work because theoretically the flame will blow off the top. Nevertheless these burners always have worked satisfactorily.

The height of the burner should be 12 metres or more above ground level and the burner should not be within 30 metres of any building, high tanks or trees, for flames 6 metres or more long are not unusual and the effects of the hot gases can be observed considerable distances away.

Waste-gas burners can be lit by sparking plugs or other special equipment, but it is always advisable to have halyards of stranded stainless steel rope by which ignited oily waste can be hoisted to the top before the gas valves are turned on from a distance. A gas burner as described, with simple means of igniting and ordinary hand-operated valves, can cost very little, but where electric and pneumatic devices of the automatic or semi-automatic push-button type have been installed the cost has been multiplied about twenty times.

15

Percolating-filter Treatment

The most important part of sewage treatment is the oxidation of those organic solids which cannot be removed by primary sedimentation, and by far the most frequent method of oxidation is aeration on percolating filters.

At the time that the Royal Commission was collecting evidence, sewage was pre-treated by chemical precipitation in quiescent tanks, chemical precipitation in continuous-flow tanks, quiescent settlement without chemicals, continuous-flow settlement without chemicals and settlement in septic tanks. The settled effluents were treated on fill-and-draw contact beds and on percolating filters. All methods of sedimentation other than continuous-flow sedimentation may now be considered obsolete as means of removing suspended solids at municipal sewage works. (The septic tank is still used at small private sewage works as will be described in Chapter 18.)

Contact beds are obsolete also. These were tanks containing a filter medium and which were alternately filled with settled sewage and emptied to aerate the medium. Aeration of the sewage took place in two or three stages. Contact beds went out of use because percolating filters could treat twice as much sewage per unit volume of filter medium and were much less liable to suffer from faulty methods of working.

Percolating filters (or trickling filters as they are called in America) are beds of medium consisting of gas-works clinker, coke, broken stone or other suitable material through which settled sewage is trickled with the aid of a distributor, usually a moving sparge pipe. After the sewage has been distributed over the medium of a new bed for some time, organic growths develop on the surface of the particles and, in the presence of the air which flows through the spaces between the particles, the organisms oxidize the greater part of the organic content of the sewage.

This organic film, which is 2 to 3 millimetres thick, is aerobic on the surface but decaying and anaerobic beneath. It is the surface film that effects the treatment and, broadly, the amount of work done by a percolating filter is in proportion to the area of film. Thus the ability of a percolating filter is in proportion to its cubic capacity for any particular grade of medium but, as a bed containing small medium has a greater area of film per cubic metre than one with coarse medium, it can effect more aeration. The smallness of medium is, however, limited by the facts that when the particles are small, the proportion of surface of particle capable of being covered with film is reduced, the flow of air through the bed is also reduced and fine media are liable to choke with solids, causing ponding and failure of the bed to perform its function.

The minimum area of organic film required to treat sewage to produce a Royal Commission effluent would appear to be in the region of 75 square metres per kilogramme $B.O.D._5$ per day in the crude sewage. In the BioDisc plant (described in Chapter 18), in which the area of film could be measured exactly, it was found that a flow of 63 cubic metres of sewage per day, loading the works with 24·8 kilogrammes of $B.O.D._5$ per day, could be reduced to a final effluent of Royal Commission standard by using a film area of 1820 square metres (or 73·5 square metres per kilogramme of $B.O.D._5$ per day). The recommendation of the Royal Commission on Sewage Disposal for the maximum quantity of sewage that could be treated on the best-quality land appears to be in reasonable agreement with this figure (see Chapter 9). If we take the areas in this recommendation and apply to them the maximum and minimum loads in kilogrammes per head per day given in Chapter 8 we have

$$2·7 \text{ m}^2 \text{ per head} \div 0·045 \text{ kg B.O.D.}_5 \text{ per head per day}$$
$$= 60 \text{ m}^2 \text{ per kg B.O.D.}_5 \text{ per day, approx.}$$

or

$$5·4 \text{ m}^2 \text{ per head} \div 0·077 \text{ kg B.O.D.}_5 \text{ per head per day}$$
$$= 70 \text{ m}^2 \text{ per kg B.O.D.}_5 \text{ per day, approx.}$$

These figures would suggest that in the best land treatment the area of land in use at any one time is completely covered with active film.

The theoretical area of the particles of medium in a percolating filter is about 82·5 square metres per cubic metre if the particles are about 38 millimetres diameter, about 70 square metres if the medium averages 45 millimetres diameter, or 40 square metres if the grade averages 78·5 millimetres, varying according to the shape and evenness of grade. But, in a percolating filter, the whole of the surface area of the particles can never be covered with film because film cannot develop and be exposed to the flow of sewage where the particles are in contact one with another, and this can mean a considerable loss of effective area. Also, a small part only of the exposed surfaces is occupied by flow unless the rate of flow is, by recirculation or other means, brought to about 5 cubic metres per day per square metre of top surface of bed.

In addition to bacteria and other oxidizing micro-organisms, various forms of vegetable life develop in the bed, including algae at the surface. These would tend to choke the medium but worms and larvae feed on the vegetable matter, causing it to break up and wash away. During the winter the algae may tend to grow excessively but there is a heavy off-loading in the spring when the worms and larvae become active.

Percolating filters are the habitat of a small fly, the psychoda and, while this does not stray far from the sewage works, it is considered one objection to percolating-filter schemes. Another cause of objection is the percolating-filter odour which, although not particularly offensive, as are septic sewage, primary sludge and screenings, is more noticeable and liable to cause complaint from the public than the odour of an activated-sludge plant.

Required properties of filter media

A material for use as a filter medium requires to be of cubical or spherical shape and not flaky, because there must be plenty of air space between the particles. It should be durable and able to withstand weather conditions and any corrosive substance in the sewage. In addition to the above necessary qualities it is advantageous for the particles to be rough so as to assist adhesion of the oxidizing organisms and to have a large surface area in proportion to volume. The material should not be expensive and, when buying filter media, it should be remembered that a large filter bed containing inexpensive material can often be more effective than a smaller bed containing some medium such as metallurgical coke which once had a high reputation but was disproportionately costly.

The grading is important. The particles should not be too small or the bed may become choked, preventing flow of air, and they must not be too large or efficiency will fall owing to lack of surface area. The author has recommended for most purposes media graded to pass a 50-millimetre sieve but to be held on a 38-millimetre sieve and, for very strong sewages, media graded from 65 to 90 millimetres diameter. These gradings are coarser than specified by some engineers in Great Britain but are in line with practice in America where, according to Imhoff and Fair [12], material that will pass a 90-millimetre sieve but will be held on a 50-millimetre sieve is commonly used.

It is of importance that there shall not be too great a difference between the sizes of the largest and smallest particles in any one bed, because small particles fill the spaces between the larger particles and, if there is too much variation of size, the flow of air will be restricted. It is also very important that the medium shall be as free as is practicable from dust of which even a small quantity can seriously interfere with performance. The danger is that, when medium is delivered to the site, dumped and rehandled into the bed, dust becomes segregated and lies in thin strata, interfering with the flow of air. It is therefore necessary for all media to be resifted after delivery to the site and thereafter handled with forks, not shovels. Care must be taken that while the bed is being filled the medium is not walked over or run over by barrows.

A percolating filter is not a filter: it is intended to pass all the suspended solids that fall upon it. The term is unfortunate for it has led some engineers to construct percolating filters as if they were filters, with media of graded sizes from fine at the surface to coarse at the bottom, and this has led to choking. Generally, percolating filters should be filled with medium of the same size throughout the depth except that, to assist the flow of air, there should be at the bottom a stratum at least 200 millimetres thick of particles of about 150 millimetres diameter

Capacities of percolating filters

To this day British practice is based on figures which appeared in the Fifth Report of the Royal Commission on Sewage Disposal [21] and which, while not described as recommendations, were expressions of opinion on the filter capacities required to effect adequate aeration. As the Royal Commission had recommended standards for treated effluents, these figures could be taken as being applicable thereto and, as the 'Royal Commission effluent' is still usually required, these figures apply as a general guide to filter capacity.

In effect, the Royal Commission suggested that an average-strength sewage, absorbing 100 to 120 milligrammes per litre of oxygen from strong permanganate in 4 hours at 27°C, could be applied to percolating filters containing coarse medium at the rate of 66 Imperial gallons per day per cubic yard of medium after 15 hours continuous-flow sedimentation and that, broadly, the rate of flow per cubic yard of medium could be varied inversely as the strength.

TABLE 46. *Appropriate loads on straight-run, coarse, deep percolating filters*

Oxygen absorbed in 4 hours (parts per 100 000)	17–25	10–12	7–8
Approximate equivalent $B.O.D._5$ (milligrammes per litre)	*804*	*421*	*287*
Gallons per day per cubic yard after 15 hours continuous-flow sedimentation	33	66	100
Cubic metres of sewage per day per cubic metre of medium	*0·196*	*0·392*	*0·594*
Cubic metres of medium per kilogramme of $B.O.D._5$ per day	*6·35*	*6·06*	*5·86*

In Table 46 are given the suggestions of the Royal Commission, in so far as they relate to straight-run, coarse, deep, percolating filters following 15 hours continuous-flow sedimentation, as compiled from the Fifth Report by the Commission's engineer, G. B. Kershaw, and, in italics, are given the appropriate loads converted to cubic metres of sewage per day per cubic metre of medium, the approximate appropriate values of $B.O.D._5$ applicable to the 4 hours of oxygen demand and, on the basis of these figures, the appropriate capacities of filter in terms of cubic metres of medium per kilogramme $B.O.D._5$ per day.

It will be noted that there is very little variation of the last figures, in spite of the considerable differences of strength of sewage, and that the average value is 6·09 cubic metres of medium per kilogramme of $B.O.D._5$ per day. If an allowance of 20% extra is made to cover the fact that present-day sedimentation-tank detention periods are in the region of 6 hours of dry-weather flow, plus other contingencies such as the effects of detergents and uncertainty as to the average ratio of B.O.D.$_5$ to the 4 hours oxygen-absorbed figure, this gives a suggested design figure of 7·3 cubic metres of medium per kilogramme of B.O.D.$_5$ per day, which is a fair representation of present-day British practice.

Applying 7·3 cubic metres per kilogramme of B.O.D.$_5$ per day to the approximate loads in kilogrammes B.O.D.$_5$ per head per day, given in Chapter 8, it would appear that, where the sewage is mainly domestic and there is a separate sewerage system, it should suffice to allow 0·4 cubic metre of medium per head of population, that where the sewage is mainly domestic but the sewerage combined 0·48 cubic metre of medium per head could be allowed and, for the average balanced domestic and industrial area with combined sewerage 0·56 cubic metre per head would probably serve. It should, of course, be understood that where an analysis of the sewage is available, this should be used in design.

Arrangement of straight-run percolating filters

For many years percolating filters were always straight-run, i.e. sewage was discharged onto them and the treated effluent ran out below: there was no recirculation of effluent or other special operation. Percolating-filter treatment is always preceded by primary sedimentation, because otherwise the beds would rapidly choke with sludge. Following the percolating filters are humus tanks to collect the suspended solids that the primary sedimentation tanks have failed to settle, plus the suspended solids produced by the filters themselves. These are usually either pyramidal-bottomed tanks or circular, mechanically-sludged tanks, and in most respects are similar to primary sedimentation tanks as described in Chapter 12; rectangular flat-bottomed tanks are obsolescent. It is usual to have a detention period of 4 hours of dry-weather flow. Scumboards are necessary because humus sometimes emits gas which lifts it to the surface.

Settled humus can be returned to the incoming flow of sewage to be resettled in the primary sedimentation tanks, can be discharged for mingling with the primary sludge, or can be passed to humus beds, which are similar in design to sludge-drying beds but are reserved for humus. The return of humus to the flow of sewage necessitates a humus-pumping station unless there is a sewage-pumping station or other pumping station at the works that can perform this duty.

The depth of a percolating filter may be limited by the available fall for, when little fall is available between the incoming sewer and the highest flood level to which the treated effluent is discharged, it is the percolating filter that can be most easily robbed of depth. A shallow filter is expensive because of the extra cost of floor and false floor. It also suffers more from freezing in the winter. Furthermore, unless there is recirculation, shallow beds are always much less efficient than those of greater depth: it is understood that, in America, tests showed that a straight-run filter 3 metres deep could be operated successfully with about eight to ten times the load that a filter 1·2 metres deep could take. As a general rule no percolating filter should be less than 1·4 metres deep from the top of the medium to the surface of floor or false floor.

It has been suggested that filters should not be too deep or there may be inadequate ventilation, but accurate information on this matter is lacking and no one has been able to confidently state what is the maximum desirable depth. On the other hand, where fall is available, increase of depth means reduction of area, with reduced cost of floor and distributing mechanisms. More important is the fact that provided the depth is not so increased as to make aeration inadequate, the reduction of surface area increases the rate of discharge onto the surface of the bed, which in turn reduces the proportion of dry patches on the particles of medium and thereby improves the performance of the bed.

It appears that the optimum performance is secured when the loading is about 5 cubic metres of sewage per day per square metre of top surface of the bed. This would require a straight-run filter to have a depth greater than has ever been allowed in practice, but high rate of discharge onto the surface can be secured by recirculation as will be described, after which depth becomes less important.

The usual maximum depth of straight-run filters is about 3 metres. The average depth of percolating filters both in Great Britain and America is in the region of 2 metres.

Recirculation and similar methods

Various experiments have been made on the operation of percolating filters. To the surprise of sewage-works managers it was found that heavy dosing of the surface at intervals produced a better result than continuous fine-spray sparging. Thus slow, creeping, distributors of rectangular beds gave better results than rapidly revolving sparge pipes on circular beds, and the performance of the circular beds could be improved by retarding the speed of rotation of the distributors.

Later there were three main methods towards which research was directed:

1. Recirculation of treated effluent through the percolating filters.
2. Alternating double filtration, which was the provision of two batteries of percolating filters, operated in series and so arranged that each, in turn, could come before the other.
3. Enclosed filtration, which was the arrangement of very deep filters covered by roofs and ventilated by forced draught.

All these methods improved filter performance, but the roof of the enclosed filter proved costly and this method has not become popular. The other two methods can be incorporated in any one plant so that effluent may be recirculated or the filters operated on alternating double filtration by adjustment of valves.

There are several ways in which a recirculation scheme can be designed. One arrangement is to collect clarified effluent from downstream of the humus tank and pump it to the flow of settled sewage from the primary sedimentation tanks to the percolating filter.

It is desirable that the rate of flow to the filter should remain constant at all times. But the rate of flow of sewage to the works varies throughout the day and from day to day, which means that the rate of recirculation of effluent should be automatically varied according to the rate of flow of sewage at the time. This can most easily be done by discharging the settled sewage and recirculated effluent to a small tank, the head in which depends on the desired rate of flow to the percolating filter. Should the flow to this tank exceed the rate which the percolating filter is intended to receive, the level in the tank will rise and the rate of recirculation of effluent will be reduced by automatic operation of an adjustable float or electrode. The maximum rate of flow to the filters need not exceed 5 cubic metres per day per square metre of bed surface, for above this value no advantage is gained. The maximum effluent-pumping capacity need not be more than this maximum rate less the peak daily flow in dry weather. There should be a sufficient number of pumps to permit a wide variation in the rate of recirculation.

While recirculation etc., can effect a great improvement in percolating-filter performance it is not at present regular practice to make any allowance for this in design.

Construction of percolating filters

Percolating filters are either circular or rectangular on plan, the shape depending on the circumstances of the site. Rectangular percolating filters are often preferred at large works or where land is restricted because they economize space. They are seldom used at small municipal works but special types are constructed at very small works for isolated buildings, as will be described in Chapter 18. The disadvantage of the rectangular filter is that it requires a heavier, more complicated distributor than does a circular bed.

The circular percolating filter can be used at works of any size, and types of distributor are available for beds of a very wide range of diameters. The main advantage of the circular bed is the simplicity of the revolving sparge by which settled sewage can be distributed over it.

The floors of percolating filters usually consist of about 150 millimetres thickness of concrete which, generally, does not need to be reinforced and requires no expansion joints. It is laid on natural ground after a sufficient thickness of surface soil has been removed. The floor of a circular filter may be arranged to fall at a slope of not less than 1 in 200 inwards towards a manhole at the centre, outwards to a peripheral collecting channel, or from one side to a collecting channel on the other. The central manhole arrangement is best if the filter has to be deep in the ground: the peripheral channel is best on a flat site when the filter can be constructed at ground level; the fall from side to side is suitable for a sloping site. With the central manhole arrangement it is advisable to have air shafts spaced around the periphery of the bed and carried down to floor level to assist ventilation.

The filter can be provided with a complete false floor of filter tiles as used at waterworks. This is the best arrangement for ensuring drainage and ventilation at the bottom of the bed, but it is expensive. In many instances it suffices to lay parallel lines of agricultural drains at not more than 1-metre centres, their ends discharging to a channel at one side of the bed, or radiating lines of tiles discharging to a central manhole or a peripheral channel. The agricultural tiles need to be set in cement to keep them in position while the bed is being filled, but the butt joints should be left open by a few millimetres to assist ventilation. On the top of the false floor or round and over the agricultural tiles is laid the stratum of coarse medium already mentioned.

The walls of the percolating filter may be constructed of brickwork, mass concrete or reinforced concrete supported on foundations at least 0·8 metre deep and strong enough to retain the filter medium. The angle of repose of gas-works clinker or coke is about 40° to the horizontal and its weight about 475 kilogrammes per cubic metre when dry. When thoroughly wet, at the optimum rate of filtration, a cubic metre of medium can contain up to 0·11 cubic metre of water and then weigh 585 kilogrammes.

The walls of a circular filter should be carried to the level of the top of the medium but not higher, as this would interfere with the proper maintenance of the sparge pipes.

Flow distribution

There should always be more than one and preferably more than two individual filters at any sewage works so that the works can continue to take the full load should it be necessary to lay off a filter for maintenance or to rest a filter if it is performing badly. To make it possible to lay off filters, the feeding arrangements, including pipework and the distributors themselves, must be of sufficient capacity for those filters that remain in service, when the maximum number has been laid off, to be able to take the peak rate of flow that the works are intended to treat in wet weather.

For the sake of economy there should not be too great a number of small filters, for large filters cost less per unit of surface area of bed than those of smaller diameter. Most manufacturers can supply distributors for circular beds of about 30 metres diameter and larger distributors have been made to order. At large works there must be many filters but a small plant can have four, the diameter being from 30 metres downwards according to the required capacity. It is usual practice, although not absolutely essential, for the flow to the various filters to be evenly distributed from a chamber which contains a separate weir and penstock to control the flow to each filter. From this chamber the feed pipe of each filter is carried below ground level, through the medium and connected to the central column. From the central column or the lowest point on the feed pipe are a pipe and sluice valve to serve as washout to the system.

The distributors of circular beds are caused to rotate by the reaction of the jets which come out laterally from the sparge pipes. At slack rates of flow this reaction may not be sufficient to cause rotation, and then the sparge pipes will stand still and dribble. This trouble should not arise at works where recirculation is continuous or at works where settled sewage is pumped to the percolating filters, and the pump can be sized to give a discharge sufficient to actuate the distributors. At other works the difficulty is overcome by the provision of dosing siphons.

Some small distributors have dosing siphons embodied in the central column but the more usual practice is to provide a siphon-operated dosing tank upstream of the weir- and penstock-chamber that distributes the flow. The siphon should be of low-draught type and the tank as shallow as practicable to economize head. The siphon should discharge, under the maximum head, the maximum rate of flow that has to be passed to the filters at any time and, when the dosing tank is nearly empty, not less than half the maximum flow. The capacity of the dosing tank should not be less than 1 minute's discharge at the maximum rate or, according to some manufacturers, between 3 and 12 cubic metres per 1000 square metres of bed surface.

The loss of head between the floor of the dosing tank and the holes in the filter arms can, in the case of small filters, be as little as 75 millimetres plus the hydraulic loss through the pipework, and the distance from the holes to the surface of the medium may be as little as 100 millimetres. But generally, and particularly for larger filters, greater head losses are involved. Where there is plenty of head available it may be used; no harm can be done by having excessive head between dosing tank and the top of the medium.

There are many different designs and qualities of rotating distributor for circular filters and, for this reason, it is not advisable to purchase the cheapest available. An engineer should recommend the purchase of the best machine that he knows and use the same type throughout the works. Good-quality distributors suffer less from corrosion and other troubles and are cheaper in the end.

The gun-metal bushed orifices in the sparge pipes are spaced so that no two shall discharge onto exactly the same part of the bed, and so that those near the periphery are closer together than those near the centre because they have to cover larger circumferences. The discharge of the sparge pipe depends not only on the loss of head down the pipe but is also influenced by the centrifugal force due to rotation. Both influences vary according to the rate of flow, and it is not practicable to calculate the exact spacings and diameters of the nozzles to ensure that each square metre of bed will receive the same rate of flow. It is therefore advisable to test distributors *under normal working conditions* by placing collecting trays on the surface of the bed and then correcting faults by reaming as necessary the gunmetal bushes until distribution is sufficiently even.

There are also many designs of travelling distributors for rectangular beds. Some are cable-driven by electric motor. Others have water-wheels actuated by the flow of sewage; it is advisable to arrange these in pairs, moving in opposite directions but connected together by cable to balance the effect of the wind, for a strong wind can oppose and bring to a halt a travelling distributor that is working on its own. Some designs of travelling distributors are driven by water-wheels carried on the sparge arms themselves. All distributors for rectangular beds are fed by siphons which suck from feed channels constructed down the centres or along the sides of the beds according to the design of the distributor.

Distributors for rectangular beds are comparatively costly and they involve losses of head from 0·6 to 0·8 metre between top-water level in the distributing channel to the surface of the medium when power-operated, and more if a water-wheel is involved.

Other forms of distribution include fixed spray jets and fixed distributor channels charged by dosing siphons. Neither of these methods is popular for municipal works in Great Britain.

16

Activated-sludge Treatment

The oxidation of sewage on percolating filters or on land is effected by micro-organisms which have attached themselves to particles of filter medium or of earth and live under a film or very shallow depth of water that flows over them. From this water they extract oxygen and this is replaced by solution from the air through a moving surface of very large area. In the activated-sludge systems oxidation is again effected by micro-organisms but they are not attached to fixed mineral matter: they cling to particles of sludge, forming floccules suspended in the turbulent contents of a tank.

Again they extract oxygen from the water and this has to be replaced but, as the liquid exposed to the air at the surface is small compared with the volume of liquid, solution at this surface is quite inadequate and replenishment of oxygen has to be effected by other means. There are two practical methods, the diffusion of air as bubbles injected at the bottom of the tank (diffused-air method) and the splashing of the liquid at the surface (surface-aeration method) which, again, forms bubbles. Both methods are effective: both require the use of mechanical power.

Choice of method depends on comparative capital costs, comparative running costs, including the effects of mechanical efficiency of aeration, and on other advantages in particular circumstances. A method that may be the best in one circumstance can be less suitable in another. For example, the diffused-air method incorporating dome diffusers has probably been most frequently chosen for large works where high-efficiency compressors can be used; but dome diffusers choke and require frequent cleaning if the sewage contains an appreciable amount of iron and, in this circumstance, either surface aeration or a coarse-bubble diffused-air method would be preferable. One reason for choosing surface aeration for the very large Crossness Sewage-treatment Works was the possibility of a large electricity demand for pumping coinciding with a power failure on the national grid. Should this happen it would be necessary to temporarily stop aeration to make available power for pumping. This would do no lasting harm to a surface-aeration plant, whereas flooding of the air mains could cause some deterioration of dome diffusers in a diffused-air plant.

In all activated-sludge works, final sedimentation tanks are provided to prevent loss of the oxidizing organisms and, in these tanks, the organisms and most other suspended solids settle readily. The sludge, known as 'activated sludge', is then returned to the incoming flow of settled sewage downstream of the primary sedimentation tanks, and the resulting mixture of settled sewage and activated sludge is known as 'mixed liquor'. The growth of the organisms is rapid and the quantity of activated sludge becomes greater than necessary, so that an excess quantity, referred to as 'surplus activated sludge', has to be withdrawn.

Just as the performance of a percolating filter depends on the exposed area of organic film, so the aeration effected in an activated-sludge plant is virtually in direct proportion to the quantity of healthy activated sludge present. This does not mean that there is a volume of aeration tank required in direct proportion to the load in terms of kilogrammes of $B.O.D._5$, for the organisms can be housed either in large aeration tanks containing a mixed-liquor of low suspended-solids content or in smaller tanks containing a proportionately denser mixed liquor but having the same weight of healthy activated-sludge solids. (The expression 'healthy activated sludge' is used because, if circumstances are adverse, in particular if the rate of aeration is too low, the sludge becomes unhealthy and the oxidizing organisms are replaced by other organisms which do not satisfactorily effect aeration.)

P. S. S. Danson and S. S. Jenkins found that the uptake of oxygen in laboratory experiments was directly related to the weight of healthy activated-sludge solids and the time during which they worked, and that this could be expressed by the formula

$$x = 22\,500yt, \tag{55}$$

where x = oxygen uptake in microgrammes,
y = activated sludge solids in millilitres,
t = detention period in hours.

The activated-sludge solids that effect aeration are not those in the aeration tanks only but all the activated-sludge organisms living in the aeration tanks, final sedimentation tanks, channels, pipes, pumping-station suction well, etc., in the circuit through which the activated sludge and mixed liquor pass. The author's examination of experimental results has shown that those organisms which are in the final sedimentation tanks are equally as active as those in the aeration tanks and any design calculations which involve the capacity of the aeration tanks only, and not the capacities of the other units in the circuit, in particular the final sedimentation tanks, lead to errors and misunderstanding. On the basis of this research he devised a formula which, converted to metric notation, reads

$$K = \frac{Q \times \sqrt{B.O.D._1} \times \log_{10}(B.O.D._1/B.O.D._2)}{78 \cdot 1 \times \sqrt{C} \times \sqrt{(kW)}}, \tag{56}$$

where Q = cubic metres per day,
$B.O.D._1$ = biochemical oxygen demand of primary sedimentation tank effluent in milligrammes per litre
$B.O.D._2$ = biochemical oxygen demand of final treated effluent in milligrammes per litre (the formula is applicable to values between the limits of 5 and 35 only),
K = a performance index dependent on efficiency of power utilization, nature of sewage, etc.,
C = Capacity of aeration plus final sedimentation tanks in cubic metres,
kW = aeration kilowatts (excluding electric power for sludge recirculation unless this is by air lift).

The figure of 78·1 is a constant inserted to give the same values of K as were obtained when this formula was expressed in Imperial units.

The formula is purely empiric, being obtained by plotting the results of numerous experiments on several kinds of logarithmic graph paper. It cannot be extrapolated beyond the limits stated. In the experiments there was little nitrification of the effluent and it is possible that an adjustment would have to be made in the case of highly nitrified effluents. However, while activated-sludge works do produce nitrified effluents, the nitrification is, on the average, much lower than that of percolating-filter effluent.

This formula explains how it is that British and American practices, in spite of their considerable differences, are both satisfactory. In the United States the average detention period of the aeration tanks is 7 hours and the average detention period of the final sedimentation tanks 2 hours, giving a total of 9 hours, whereas in Great Britain the average detention period of the aeration plus final sedimentation tanks is 12 hours dry-weather flow or $1\frac{1}{3}$ times the United States figure. On the other hand, the average free air input in the latest activated-sludge plants in Great Britain is 47 cubic metres of air per kilogramme of $B.O.D._5$ to be removed from the settled sewage, whereas the Americans allow 62·5 or $1\frac{1}{3}$ times the British figure. Thus, the product of detention period and air input is the same on both sides of the Atlantic and, as will be seen, is in conformity with the formula.

It is common practice to design activated-sludge plants to secure final effluents having a $B.O.D._5$ value of 15 milligrammes per litre so that the effluent will seldom, if ever, have a $B.O.D._5$ of more than 20 milligrammes per litre, and higher standards are often obtained.

Diffused-air system

There are several varieties of the diffused-air system differing in the type and position of diffuser. In Great Britain the diffused-air system almost invariably means the method in which air is introduced at the bottom of aeration tanks via porous ceramic diffusers, the purpose of which is to ensure that the bubbles are small and the materials for which are supplied by the old-established firm of Activated Sludge Ltd. This method has been undergoing modifications in view of experience gained ever since it was first introduced prior to the First World War. At one time various arrangements were tried, including floors with individual pockets, transverse ridges and furrows, longitudinal ridges and furrows and spiral flow. Of these the longitudinal ridge-and-furrow arrangement was mostly used, in which aerating tiles were placed at the bottoms of furrows in the concrete floors. In the spiral-flow method the air diffusers were placed near the bottoms of the tanks at one side only so that the mixed-liquor flowed spirally from inlet to outlet, the cross-velocity preventing settlement of sludge. (In all methods it is essential that the sludge shall not settle, for if it does, it turns septic and the quality of the effluent suffers.)

At the present time, aeration tanks for the diffused-air method of Activated Sludge Ltd., are flat-bottomed and much shallower than formerly. The tanks can be from 2 to 3 metres deep, 2·5 metres being most usual, and of any width, not necessarily multiples of some standard figure. There should not be fewer than four individual aeration tanks at any works.

The diffusers are the kind known as 'dome' diffusers. These are 178-millimetre diameter mushroom-shaped porous tiles secured by non-ferrous screws to plates, and sealed with rubber rings to prevent air from escaping other than through the diffusers. The diffusers are fixed to special PVC air mains that are arranged along the bottoms of the aeration channels. Each diffuser will pass from 0·015 to 0·03 cubic metre of air per minute according to the head. A discharge of 0·018 is considered the approximate working figure and it is generally preferred that the maximum discharge should not be more than 0·023. The loss of air pressure through the diffuser when the discharge is 0·018 cubic metre per minute is the equivalent of 0·225 metre head of water when the diffusers are clean and new.

It is usual in British design to allow a maximum air flow of 15 cubic metres of free air per cubic metre of sewage treated where the sewage is mainly domestic and on the separate system, about 18 where the sewage is mainly domestic and on the combined system and, for the average balanced domestic and industrial area on the combined system, about 21. Another rule is to allow, for average working, 47 cubic metres of free air per kilogramme of $B.O.D._5$ to be removed from the settled sewage to produce the required standard of effluent and to add a factor of safety of 67% for maximum conditions. In practice the amount of air used is less than the design figure and varies between 9 and 12 cubic metres per cubic metre of sewage treated.

Detention period

From what has been said it will be observed that, to conform with either British or American average practice, the combined detention periods of the aeration tanks, final sedimentation tanks and any sludge reactivation tanks or aerated-sludge or mixed-liquor delivery channels could be calculated by the formula

$$T = 564/A, \qquad (57)$$

where T = aggregate detention period of the aforementioned units in hours dry-weather flow,

A = air supply in cubic metres free air per kilogramme of B.O.D.$_5$ in the settled sewage delivered to the aeration tank.*

This detention period can be shared between the units in reasonable proportions. In Great Britain the detention period in the final sedimentation tanks seldom exceeds 6 hours and is seldom less than 4 hours. These periods, which are long compared with those used in America, are preferred because a large final sedimentation-tank capacity acts as a reservoir for sludge during rapid changes in the rate of flow of sewage.

Virtually the whole of the remainder of the aggregate detention capacity can be in the aeration tanks but, formerly, British practice was to reserve 10% to 20% of the aeration channels for 'reactivation' of activated sludge, i.e. aeration of sludge withdrawn from the final sedimentation tanks prior to mingling it with settled sewage.

It is convenient to connect the aeration channels to the final sedimentation tanks or to return the activated sludge withdrawn from the final sedimentation tanks to the inlet ends of the aeration tanks via open channels aerated in the same manner as the aeration tanks. This makes it possible for the channels to flow at low velocities, economizing head. The channels must be the same depth as the aeration tanks so that the same air pressures can apply. Capacities of all these aerated channels, as has been said, should be included in the aggregate detention period.

The surface area of the aeration tanks can be determined by the space required to accommodate the number of diffusers necessary to pass the maximum calculated quantity of free air. For example, if diffusers are spaced 0·3 metre apart along air mains which are laid in the bottom of the tank at 0·75 metre centres, the normal design rate of air flow will be 0·08 cubic metre per minute per square metre of surface of tank. This could be a reasonable figure for aeration tanks 2·5 metres deep. With shallower tanks or for stronger sewages, the spaces between the lines of air mains could be reduced to 0·6 metre.

Air supply

Air must be capable of delivery from the compressors at the maxi-

* Average working supply, not the maximum design figure which should be about 1·67 times higher.

mum quantity and at the pressure required to overcome the loss of head through the diffusers, the resistance of the depth of water and the loss through the pipework. The diffusers, mounted on their supply pipes, are about 0·225 metre above the floor of the tank and, to allow for their discharge becoming reduced by dirt, it is usual to assume that the air has to be delivered from water level to the bottom of the tank or against a pressure of 0·1 kilogramme per square centimetre for every metre depth of water.

It is important that there should not be too great a difference of pressure on those diffusers which are near the compressors as compared with those which are far away. For this reason the air-supply pipework used at activated-sludge works is much larger in diameter

TABLE 47. *Recommended sizes of air mains for diffused-air works*

Internal diameter (nominal millimetres)	Free air discharge (cubic metres per minute)
100	3·75
150	11·05
225	32·2
300	69·7
375	126
450	204
525	311
600	437
675	601
750	802
825	1030
900	1300
975	1580
1050	1960

than is usual in other compressed-air practice. Table 47 gives recommended diameters for air mains which generally will not give a loss of head of more than 0·063 metre per 100-metre length of pipe when the pressure at the compressor is about 2 metres or 0·058 when the pressure at the compressor is about 3 metres. (The loss varies inversely as the ratio of absolute pressure in pipes at entrance to atmospheric pressure.) Air mains should be provided with blow-out pipes at the end of each section both for the removal of condensation water or oil, and also in case the mains should become filled with water during a failure of air pressure. There should be a silencer between the compressor and the air mains because compressors can be noisy.

The air should be compressed by rotary blowers and not by reciprocating compressors because the latter require internal lubrication which could lead to chokage of the diffusers. There must be a sufficient number of compressors to allow for stand-by and to facilitate variation in rate of flow of air but, whenever practicable, the larger sizes should be used because of the appreciably greater efficiency (see Table 24).

It is necessary for the air to be filtered before delivery to the air mains because removal of dust greatly increases the life of the diffusers which, without filtration, will eventually choke and have to be cleansed by reburning or treatment with chemicals. Electrostatic filters are the most effective and are used at large works. For small plants less expensive filters, consisting of mats of metal, glass wool or hair, covered with viscous oil, are used. (For compression of air see Table 21.)

Recirculation of activated sludge

In present-day British practice it is usual to provide for the possibility of extracting activated sludge from the final sedimentation tanks and returning it to the inlet ends of the aeration tanks in a quantity equal to the design dry-weather flow but to make it possible for the rate of recirculation to be reduced to two-thirds or one-third as necessary. Recirculation can be effected by axial-flow pumps, which are usual in surface-aeration schemes: low-lift air lifts (see Chapter 4) are usual for diffused-air plants.

Surplus activated sludge

From the flow of returned activated sludge is withdrawn, as required, the surplus activated sludge in instrumentally-measured quantity. The amount of this can vary considerably: it averages 3·88 cubic metres per 1000 head of population per day at an average moisture content of 99·33% (see Chapter 13). The arrangement of instruments, pipework and valves should allow for the steady withdrawal of surplus activated sludge at a minimum of about one-third of this average rate. As it might be necessary to make a rapid evacuation should there be serious bulking or other break-down of the treatment, it should also be possible to draw off at about one-quarter of the maximum rate of *recirculation*. The surplus activated sludge should be returned to the incoming flow of sewage for resettlement in the primary sedimentation tanks or passed to an activated-sludge thickening plant such as the Dissolved Air Thickening Process (see Chapter 13).

INKA system

The INKA system is a diffused-air method developed in Sweden and installed at several works in Europe. In lieu of ceramic diffusers, aerators of stainless steel or plastics materials, in the form of a grid with 2·5-millimetre diameter holes on the underside, are arranged along one side of a spiral-flow tank at a depth of 0·6 to 0·75 metre. The tanks are 3 to 3·5 metres deep and 3 to 9 metres wide. The grid systems vary from 2 metres wide for tanks up to 6 metres wide and 2·25 metres wide for tanks up to 9 metres wide. A vertical fibreglass baffle is provided to assist the spiral flow. Air is blown into the system by centrifugal fans at the pressure necessary according to depth of aerators etc., and in the quantity of about 33 cubic metres per cubic metre of sewage treated. This process is marketed in Great Britain by Dorr–Oliver Co., Ltd.

Mechanical agitation or surface aeration

The surface-aeration methods are broadly the same as the diffused-air methods, involving the use of aeration tanks, final sedimentation tanks and recirculation of activated sludge. The main difference is that the aeration is produced, not by blowing air through the mixed liquor but by causing splashing at the surface, which has the same effect.

Originally the aeration tanks had detention periods in the region of 16 hours of dry-weather flow because, at that time, the methods of mechanical aeration did not cause sufficient aeration to permit detention periods as short as those applicable to the diffused-air system. This difficulty has now been overcome by the improved and ever-improving design of aerators and, at the present time, the detention periods used are very similar to those applicable to the diffused-air system at similar power demands.

Of the surface-aeration methods, by far the most popular in Great Britain is the 'Simplex' system of Ames Crosta Mills & Co., Ltd. Originally this involved tanks with a series of hopper bottoms. At the centre of each hopper bottom was an uptake tube at the top of which was a mixed-flow impeller, described as an 'aerating cone'. This cone drew the mixed liquor up the tube and flung it over the surface of the mixed liquor in the tank.

With increase of power input and the reduction of tank capacity, the cones became known as 'high-intensity' cones. These have been greatly improved in efficiency. Owing to the very high degree of turbulence the tanks no longer need to have hopper bottoms, but are flat-bottomed merely with chamfers at the edges.

The tanks are provided with adjustable-level outlet-weir channels the purpose of which is to control the submergence of the cones. Increase of submergence by a maximum of 75 millimetres makes a small increase of power demand and proportional addition in aeration.

The 'Simplex' plant designed by the author for the Crossness Sewage-treatment Works, which at the time of writing is believed to be the largest surface-aeration plant in the world, has hopper bottoms because of its date of installation, although the present rate of aeration is high enough to justify flat floors. The most recently published figures for performance (1st April to 30th September 1968) showed that this plant dealt with an average daily flow of 554 545 cubic metres of sewage from a population of 1 700 000. The power input was 2630 kilowatts, the capacity of aeration tanks in use at the time was 131 000 cubic metres and the installed capacity of the final sedimentation tanks was 104 000 cubic metres, giving a total of 235 000 cubic metres capacity in aeration and final sedimentation tanks. (No information was available on any final sedimentation tanks which may have been laid off at the time.) The biological chemical oxygen demand of the settled sewage was 148 milligrammes per litre and that of the final effluent 13·5. (This was according to the standard method of analysis: the G.L.C. uses another method also which is modified to prevent nitrification of the sample during the test and gives lower values for final-effluent $B.O.D._5$ than are obtained by the standard method.) According to Formula 56 the value of K was approximately 3·6.

The above are typical figures for a 'Simplex' scheme in which the type of high-intensity aerating cone available at that time is used. They could safely be used as a guide to design.

Kessener system

The Kessener system, introduced by Dr Kessener of the Netherlands, involves a spiral-flow tank along one side of which is a rotating brush or series of stainless steel combs which spray the mixed-liquor over the surface. The tank is specially shaped and provided with a baffle board to assist spiral flow.

Bio-aeration

Mention should be made of the Sheffield or 'bio-aeration'* system because, although it is tending towards obsolescence, many works of this type are still in operation in Great Britain. This was the first type of surface-aeration plant to be used. The aeration tank is a long narrow channel about 1·4 metres wide and deep and having a detention period of about 16 hours. The channel is folded upon itself again and again. Across the folded channels are line shafts driving paddle wheels which cause surface agitation and propel the mixed liquor at a velocity of about 0·5 metre per second. Activated sludge, collected in the final sedimentation tanks, is mixed with the settled sewage at the inlet end.

Tapered aeration and incremental feeding

Tapered aeration is the name given to the application of more air at the inlet end of a diffused-air aeration tank and less at the outlet end, the proportions being 45% of the total air in the first third of the tank, 30% in the second third and 25% in the last third. Incremental feeding or step-aeration is the form of operation in which activated sludge is introduced at the inlet end of any type of aeration tank and sewage is added later at various points along the length of the tank. This is considered by some to be a substitute for reactivation of sludge in separate tanks. Tapered aeration (first installed at Worcester

---

* The term should not be confused with 'bio-filtration', the name applied to the recirculation methods preferred by Dorr-Oliver Co., Ltd, or 'bio-flocculation' which means activated-sludge treatment preceding percolating-filter treatment.

in 1916) has fallen into disuse but several engineers adhere to the practice of incremental feeding and, at present, it is usual to provide the channels and penstocks necessary to make it possible. However, the lateral velocities produced by present-day high-rate aeration must result in complete mixing of activated sludge and sewage throughout the length of an aeration channel in a matter of a few minutes,* no matter how far apart the inlets for sludge and sewage may be and, as detention periods are measured in hours, it would appear impossible for either tapered aeration or incremental feeding to have any effect whatsoever.

Bulking

The activated-sludge methods require intelligent operation, involving study of laboratory tests, for they are more liable to trouble than percolating-filter treatment. It is important that a suitable proportion of activated sludge should be retained to preserve the correct proportion of solids in the mixed liquor according to the strength of sewage and the detention period. Also it is necessary for the rate of aeration by diffusion or otherwise to be adequate. If the load on the works becomes too great for the organisms or for the amount of air provided to them, a change known as bulking takes place and other organisms, including filamentous growths and types of protozoa, replace the oxidizing organisms. Then the sludge becomes thin and difficult to settle and the quality of the effluent deteriorates. To avoid bulking, the load on the works should not exceed 0·3 kilogramme B.O.D.$_5$ per day per kilogramme of volatile solids in the aeration tanks etc., and the air supply should be adequate.

Sludge-volume index

The sludge-volume index is used as a means of ascertaining the condition of the activated sludge. To obtain it, a sample of mixed-liquor is taken from the aeration tanks and allowed to settle in a 1-litre graduated cylinder, when the percentage volume occupied by the settled sludge is recorded. At the same time the suspended-solids content of another portion of the original sample is determined. The sludge-volume index is then found by the formula

---

* In one test lateral diffusion was at 0·2 metre per second.

$$\text{Sludge-volume index} = \frac{\%\text{ settling by volume in 30 min}}{\%\text{ suspended solids}} \qquad (58)$$

A good sludge should have an index of not more than 100.

Sludge age

Another method of determining the condition of activated sludge is finding the sludge age in days. There is more than one way of determining this; perhaps the best is by using the following formula

$$N = TQ_1/24Q_2, \qquad (59)$$

where N = sludge age in days,
Q_1 = daily quantity of circulated activated sludge,
T = detention period in aeration, final sedimentation tanks, etc., in hours,
Q_2 = daily quantity of surplus activated sludge.

A difficulty when comparing the results of different works is that this and other formulae usually have been applied without allowance being made for the time the activated sludge has been retained in the final sedimentation tanks, which must contribute to the actual age.

Foaming

Increased use of synthetic detergents has made sewage very liable to foam. This is particularly noticeable in aeration tanks where, if no precautions are taken, foam may rise to a height of several feet and be blown for considerable distances outside the boundaries of the works. This foam contains bacteria and objectionable substances. When it dries, it leaves a black deposit. Consequently it can be a just cause of objection from nearby property occupiers.

Of the methods that have been tried for reducing foam is the application of suitable oil upstream of the aeration tanks or other places where foaming may be liable to occur. The quantities required vary between 1·5 to 3·5 milligrammes per litre of sewage. It can be applied at various points by pumping with the aid of mechanical lubricators, of the kind used in large internal combustion engines, through small-diameter tubing. For this purpose nylon or PVC tubes of 0·32 to 0·64 millimetres diameter have been used. These should be in complete lengths without joints from the lubricator to the points of discharge.

The oil is stored in tanks and delivered to ball-valve controlled cisterns from which the lubricators pump. Remote control of the mechanical distributors is desirable in order that only those which are required may be brought into operation. Fortunately the detergents now in use are much less liable to cause foaming than those used some years ago.

17

Production of Effluents of Extra High Standard

Now that river boards may require very high standards of final effluent, in circumstances that justify the expense (see Chapter 8) special methods of complying with these requirements are coming into use. As has been mentioned, it is possible to produce superlative effluents by having treatment works of extra large size, but this is not an economic proposition for the additional cost is out of proportion to the advantage gained. Suppose, for example, that an activated-sludge plant were to be designed to treat 25 000 cubic metres per day of settled sewage that had a $B.O.D._5$ value of 156 milligrammes per litre, that a final effluent $B.O.D._5$ of 15 was required to ensure that the $B.O.D._5$ seldom exceeded 20, that the value of K for the type of works was 3·55 and that there was an aeration power input of 350 kilowatts, it would be found by Formula 56 that the condition could be satisfied by having a capacity of 12 500 cubic metres in the aeration tanks and final sedimentation tanks. If, however, the river board called for a final effluent $B.O.D._5$ of 10 milligrammes per litre, which would mean designing to the figure of 7·5, the new requirement could be satisfied by having a capacity of 16 750 cubic metres and a power input of 470 kilowatts: in other words, both the capital cost and the running costs of the aeration plant, including final sedimentation tanks, would have to be increased by about one-third.

The same result could be obtained at much less cost by filtering the final effluent with filters of sufficiently fine grade to reduce the suspended-solids content by about 23 milligrammes per litre. This should secure the desired effect, for the removal of suspended solids from a final effluent reduces the $B.O.D._5$ also by about one-third of the amount of the suspended-solids reduction (see Chapter 8).

This type of treatment is known as effluent polishing. The methods used include various types of filtration as normally employed at waterworks or, alternatively, some forms of land treatment. When an effluent is filtered, this is always after adequate settlement in humus tanks, in the case of a percolating-filter scheme or, of course, in final sedimentation tanks in an activated-sludge scheme, because sedimentation greatly reduces the load on the filters and therefore the size of plant required.

Glenfield & Kennedy manufacture Micro-strainers which are revolving-drum filters, the filter medium of which is a fine stainless-steel gauze. The gauze is cleansed by backwashing with treated effluent, the wash-water being returned upstream of the strainer. From time to time the fabric has to be cleaned to remove organic growths which are not inhibited by backwashing. Micro-strainers with a gauze of suitable grade will reduce the solids-content of an average humus-tank effluent from 30 to about 10 milligrammes per litre.

Rapid sand filters, as used at waterworks, have been found capable of treating 165 cubic metres of humus-tank effluent per day per square metre of surface, under a maximum head of 2·75 metres. When the loss of head had reached as much as 2 metres, which happened every 8 to 10 hours, the filters were backwashed with filtrate at a rate equal to 9·5% of the flow. This filtration reduced the suspended-solids content of the effluent to 4·6 milligrammes per litre. For design purposes not more than 100 cubic metres per square metre per day should be allowed.

Rapid filters are similar to slow sand filters in that the liquid is filtered through a bed of sand but, in other respects, construction and mode of operation differ. The filter medium is a sand of 0·4- to 0·8-millimetre grade and can be 0·75-metre deep. This should rest on a bed of gravel 0·15- to 0·45-metre deep, graded from 2·5- to 50-millimetre diameter particles from the top to the bottom. The filtered effluent is collected by perforated pipes below the gravel, floor channels with perforated plates or a false floor of filter tiles. The filtrate for backwashing is introduced through the perforated pipes or plates in quantities sufficient to give an upward velocity of 0·6 metre per minute. The overflow of this wash-water is collected in troughs placed well above the surface of the sand.

Slow sand filters of the waterworks type are not recommended for sewage effluents but can be used. These consist of beds of sand of about 0·3 millimetres effective size laid to a depth of 0·8 metre below which is 0·225 metre of graded gravel supported on filter tiles or porous slabs. (By effective size is meant the size of the largest particles in the finest 10% of the whole.) The depth of water over the top of the sand varies but, for design purposes, should be taken as a maximum of 1·2 metres. The beds are rectangular. An apron is provided at the inlet to prevent disturbance of the medium by the inflow and the inlet chamber has a washout for draining away top waste. It is not usual to backwash slow sand filters and, when the head exceeds about 0·6 metre, they are cleansed by being drained via the top-waste valve, after which the sludge is scraped off and such surface sand as is lost in this process is replaced with new.

Slow sand filters (when used for water) will pass about 2·44 cubic metres per square metre of surface per day.

Treatment on land

Settled effluent can be given further treatment by land filtration, as described in Chapter 9, where the subsoil is of suitable quality and land is available. It has the advantage of requiring less maintenance than filters. Short-period sedimentation is advisable after land treatment to remove any suspended solids drained from the land.

Broad irrigation over ploughed land has not been favoured as a means of effluent polishing, but irrigation of humus-tank effluent over grass plots has been tried at the works of the Birmingham Thame and Rea District Drainage Board. An area of about 1 square metre per head of population was allowed. The area was divided into a number of plots to facilitate maintenance and reconditioning, each plot being about 90 metres long from inlet to outlet on land falling at 1 in 60. Maintenance consisted of drying out the plot and mowing twice in the growing season to discourage weeds. The effect of this treatment was to reduce the suspended-solids content of a humus-tank effluent from 17·0 to 8·1 milligrammes per litre and the $B.O.D._5$ from 16·2 to 7·6 milligrammes per litre.

The author has used irrigation over grass land in lieu of humus-tank settlement at small works for institutions and has obtained very high-quality effluents. The area allowed was in the region of 3·34 square metres per person served.

Algae Ponds

A method by which the dissolved-solids content of treated effluent can be reduced and nitrates can be salved is the use of algae ponds. In these ponds algea are encouraged to grow and are then harvested in healthy condition before decay sets in. The ponds are drained in turn, the algae scraped from the sides and bottoms, and then the ponds returned to use. The method effects some reduction of $B.O.D._5$ because of the oxygen given off by the algae, and removes bacteria to some extent. A particular advantage of the use of algae ponds is that it reduces the growth of algae in the stream to which the final effluent is discharged: such growth can otherwise become excessive where a well-nitrified effluent has been obtained.

18

Sewage Treatment for Isolated Buildings

Wherever it will not be unreasonably costly, premises should be connected to sewers draining to municipal works, because this is in the interests of public health. Also it is often less expensive than properly constructed private sewage-treatment works, even in many localities where properties are not close together: this is a fact that is seldom appreciated. In more than one instance the author has found it economical to lay a gravity drain a mile or more in length or to pump a quarter of a mile to connect to a public sewer in lieu of building or reconstructing private sewage-treatment works. Properly constructed small sewage-treatment works are expensive, as will be observed by referring to Formula 43 which, at the time it was devised, was applicable to works serving single houses or bungalows as well as the largest municipal works.

The present position in Great Britain is far from satisfactory for, under the guidance of ill-informed contractors or architects and even official publications, property owners or occupiers have been persuaded to agreed to the installation of 'sanitation' that amounted to little more than a pretence and that involved risk of causing nuisance and danger to health. In the examination of the sewage-disposal arrangements of twenty-eight premises in the ownership of one body, the author found nine cesspools unlawfully overflowing to streams, ditches or lakes, five cesspools unlawfully discharging to soakaways, one cesspool discharging on to the surface of the land with no proper irrigation, one cesspool in the form of a simple excavation from which the effluent unlawfully soaked away, twelve sewage-treatment works involving percolating filters in all states of disrepair from completely derelict, and all giving very unsatisfactory effluents. There was not one properly constructed and operated cesspool and not one sewage-treatment plant properly maintained and giving a reasonable effluent.

SEWAGE TREATMENT FOR ISOLATED BUILDINGS

This sample survey is typical of what is to be found all over the country, and this state of affairs is inexcusable. It is due to ignorance and irresponsibility, for it is always possible to design, construct and maintain small sewage-treatment works capable of producing effluents equal in quality to those expected at municipal works.

The methods of sewage disposal for isolated buildings or small groups of buildings include the following:

1. Connexion by gravity or pumping to municipal sewers. (The possibility of this should always be considered first and comparative estimates made between costs of main drainage and those of *properly constructed and maintained* local works.)
2. Conservancy sanitation.
3. Local sewage-treatment plant.

Conservancy sanitation

Conservancy sanitation is the storage of sewage until it can be removed completely and in a sanitary manner to some place of disposal such as a sewer manhole or the sewage-treatment works of a local authority. It includes the use of earth-closets and cesspools. The Public Health Act, 1936, permits conservancy sanitation as will be seen from the following quotations:

'37—(3) A proposed drain shall not be deemed to be a satisfactory drain for the purposes of this section unless it is proposed to be made, as the local authority, or on appeal a court of summary jurisdiction, may require, either to connect with a sewer, or to discharge into a cesspool or into some other place:

Provided that, subject to the provisions of the next succeeding subsection, a drain shall not be required to be made to connect with a sewer unless—

(a) that sewer is within one hundred feet of the site of the building or, in the case of an extension, the site either of the extension or of the original building, and is at a level which makes it reasonably practicable to construct a drain to communicate therewith, and, if it is not a public sewer, is a sewer which the person constructing the drain is entitled to use; and
(b) the intervening land is land through which that person is entitled to construct a drain.'

'43.—(1) Where plans of a building or of an extension of a building are, in accordance with building byelaws,* deposited with a local authority, the authority shall reject the plans unless either the plans show that sufficient and satisfactory closet accommodation consisting of one or more waterclosets or earthclosets, as the authority may approve, will be provided, or the authority are satisfied that in the case of the particular building or extension they may properly dispense with the provision of closet accommodation:

Provided that—
(i) unless a sufficient water supply and sewer are available, the authority shall not reject the plans on the ground that the proposed accommodation consists of or includes an earth-closet or earthclosets; and
(ii) if the plans show that the proposed building or, as the case may be, extension is likely to be used as a factory, workshop or workplace in which persons of both sexes will be employed, or will be in attendance, the authority shall reject the plans, unless either the plans show that sufficient and satisfactory separate closet accommodation for persons of each sex will be provided, or the authority are satisfied that in the circumstances of the particular case they may properly dispense with the provision of such separate accommodation.'

Chemical closets

At the present time chemical closets are largely used in lieu of earth-closets, and they may be so used because, legally, they *are* earth-closets, provided that they conform to the definition of earth-closet in the Act:

'earthcloset' means a closet having a *moveable*† receptacle for the reception of faecal matter and its deodorisation by the use of earth, ashes or chemicals, or by other methods,

and with Building Regulations which, broadly, require that any earth-closet which is not a chemical closet shall be so constructed that it can be entered from the external air only or from a room or space which itself can be entered directly from the external air only.

* Now replaced by Building Regulations (author).
† Author's italics.

No earth-closet or chemical closet shall open directly into a habitable room, kitchen or scullery or room in which any person is habitually employed in manufacture or business. An earth-closet which can be entered directly from the external air shall have an adequate ventilation opening as near the ceiling as practicable and one which cannot be entered from the external air shall have a window or similar means of ventilation to the external air capable of opening to not less than one-twentieth of the floor area. Earth-closets shall be so situated that they shall not be liable to pollute any source of water used or likely to be used for domestic purposes.

The Regulations also require that the floor of the closet shall be non-absorbent and, if the closet is entered from the external air, shall be not less than 75 millimetres above the adjoining ground and fall towards the entrance at not less than 1 in 24. The receptacle shall be non-absorbent and so constructed and located that the contents *cannot escape by leakage or otherwise* or be exposed to rainfall or other drainage. The receptacle and fitting shall be so constructed and arranged that maintenance shall not constitute a nuisance or be a danger to health. *No part of the receptacle of an earth-closet or chemical closet shall discharge to a drain.*

Warning should be given that, in addition to some available patterns of chemical closet that comply with statute and regulation, there appear on the market from time to time various designs that are not 'closets' within the meaning of the Act and which offend against Building Regulations and the interests of public health.

Where chemical closets or other earth-closets are installed, there should be proper arrangements for disposal of their contents, preferably by the local authority. Many local authorities perform this service and there are special arrangements on some cesspool-emptying vehicles to facilitate closet-emptying and cleansing.

It is necessary to have sufficient closet capacity to store night-soil between the visits of the local authority's vehicle, due allowance being made for the possibilities of a delay in the service or of overloading of the closets by visitors. This means that the smallest premises should have no fewer than two chemical closets installed. It is true that chemical-closet contents have been used as manure or have been discharged on to land as a means of disposal, but this is not desirable in all circumstances; it has been reported that the chemicals of the kinds used in chemical closets have been detected in public water supplies.

The following figures should be useful in estimating closet requirements. The average amount of night-soil is 15 litres per head per week, to which should be added about 4·5 litres per week for chemicals and dilution water per closet. As the effective capacity of chemical closets, as manufactured, usually does not exceed 25 litres, a closet which is emptied once a week is hardly adequate for two persons. But, in spite of the above figures, a chemical closet will usually be adequate for a household of two persons, if emptied once a week: this is no doubt due to the fact that all members of a household are not at home all the time.

Three types of chemicals are used in chemical closets:

1. Coal tar preparations with oil seal.
2. Coal tar preparations without seal.
3. Formaldehyde preparations.

The oil seal in the first of these mixtures is intended to prevent evaporation from the surface of the contents with the possibility of bad odours. The oil must be of a type that will not catch fire should a match or burning paper be dropped in the closet and it must readily separate from the fluid to form a surface layer. When a preparation containing oil is used, the chemical must be shaken before application.

The remaining chemicals are required to inhibit the action of bacteria. They must be strong enough to remain effective with the maximum dilution they are likely to receive and for a period of not less than one week. They must not have any objectionable odour themselves. They must not be injurious to the skin or clothing and they must not be liable to render cleansing of the closet difficult. They should not be liable to damage any of the materials of which the closet is made: in this connexion it is advisable to use the chemical supplied by the maker of the closet. The chemicals are usually coloured to obscure the contents of the closet.

Sullage disposal

Where chemical closets are used for the reception of night-soil there remains the problem of disposing of sullage from kitchen sinks, clothes washing, slops, etc. These are almost invariably discharged to pits which are in fact cesspools that do not conform to the law. From these the contents overflow or soak into the soil and, while sullage is not so objectionable as night-soil or domestic sewage containing night-soil, it can be dangerous and has been the cause of many deaths by contamination of water used for domestic purposes. This is a fact which local authorities often overlook.

All sullage should be discharged to proper cesspools designed and constructed in accordance with the Public Health Act, 1936, and Building Regulations and should be regularly emptied by the local authority or a contractor. Chemical closets are used because of inadequate water supply; it follows that, where chemical closets are used, the quantity of sullage is not likely to be great, and therefore a sullage cesspool will need emptying much less frequently than a cesspool taking flows of sewage from houses which, having adequate water supply, have water-closets, sinks, lavatory basins and baths.

Cesspools

The Public Health Act, 1936, defines a cesspool as a settlement tank or other tank for the reception or disposal of foul matter from buildings. The term covers two very different things:

1. An absolutely watertight storage tank that has to be emptied from time to time in a sanitary manner.
2. A sedimentation tank the effluent of which passes *to some proper form of secondary treatment* and from which the sludge only has to be removed in a sanitary manner.

The Act empowers the local authority to require any person who has caused or permitted soakage or overflow from a cesspool to take such steps as may be necessary to prevent the soakage or overflow, except where the cesspool is a properly constructed tank for the treatment of sewage and the effluent is conveyed away in such a manner as not to be prejudicial to health or cause a nuisance.

The Building Regulations, 1965, require that any cesspool, including a settlement tank, septic tank or other tank for the reception of foul matter from a building shall be so constructed as to be impervious to water both from inside and outside. It shall be sited so as not to render any spring stream or well used or likely to be used for domestic purposes liable to pollution. There shall be means of access for cleansing without the need for carrying the contents through any building used for residence or trade or to which the public has access. The cesspool shall not be so close to a building used for residential or business purposes or to which the public has access as to be liable to become a nuisance or danger to health.

Building Regulations require that any cesspool not a settlement or septic tank (which presumably means a tank used as part of a treatment plant) shall be of a depth that permits complete emptying, covered to exclude surface-water or rainwater, fitted with a suitable manhole cover, adequately ventilated and without any outlet for overflow or discharge. Such a tank must have a capacity below the level of the inlet pipe of not less than about 18·2 cubic metres. This last is a new requirement and is a step in the right direction for it will, in many instances, make cesspools more expensive to construct than proper sewage-treatment works and discourage their use.

Any settlement tank or septic tank that comes before secondary treatment shall be of suitable depth and adequate size for the purpose and in no case of a capacity less than 2·73 cubic metres. It shall be covered and adequately ventilated with means of access for inspection and cleansing or, if not covered, it shall be fenced in.

The regulations that cesspools shall be of the depth necessary to enable complete emptying and of capacity not less than 18·2 cubic metres virtually determine the design of those of minimum size. The depth is limited to 5 metres below the level of the adjacent road which will often mean an effective storage depth of no more than 4 metres. This will mean that the diameter (for most cesspools are circular on plan) must not be less than 2·4 metres.

Because the name 'septic tank' gives confidence, house agents and others concerned with property often state that a septic tank is provided, where, in fact, there is only a cesspool. While, in legal definition, a septic tank is a cesspool, a cesspool intended for storage of sewage is not a septic tank; the latter is a settlement tank that *must* precede some form of secondary treatment.

Cesspools are emptied by tank-vehicle using pneumatic suction. This may be a local authority service or, in the absence of such service, the work may be done by contractor. The emptying should always be supervised or it may become little more than a gesture. Cesspool-emptying vehicles have tanks of various sizes, the most usual being in the region of 3·4 cubic metres. It follows that a cesspool constructed to present-day regulations *cannot* be emptied in fewer than six journeys, particularly as the vehicles are often observed to go away incompletely filled. If a cesspool is said to be emptied in a single journey, this means either that the work has been scamped or that the cesspool is leaking seriously.

Many existing cesspools are emptied by chain pump for the disposal of the sludge onto the land as manure. This is permissible only in very out-of-the-way places, for the procedure of discharging the contents of the cesspool is most offensive and discharge on to the land can be permitted only where there is no risk of nuisance or contamination of water supplies. Chain pumps can be hand- or power-operated. They consist of an endless chain carrying discs which transport liquid and solids up a vertical pipe that discharges about 1 metre above the base plate (or ground level). They are available to draw from depths between the limits of about 3·7 and 7·6 metres. (For capacities see Table 48).

Small sewage-treatment works

All premises having their own private sewage-treatment plant should be drained on the strictly separate system. Rainwater in large quantity interferes with sedimentation and the performance of percolating filters, and on very small sites the ratio of rainfall run-off to flow of soil sewage can be very large indeed. When new treatment works have to be provided for existing premises that already have a combined drainage system, the designer has to consider the advisability of reconstructing the drains as an alternative to a very difficult problem of separation and disposal of storm water.

Methods of sewage treatment for isolated premises include:

1. Treatment by settlement tank and percolating filters.
2. Treatment by settlement followed by activated-sludge aeration.
3. Treatment by settlement tank and land filtration or broad irrigation.

The method to be chosen depends on circumstances but sedimentation followed by percolating-filter treatment is by far the most usual because it occupies little space, will work satisfactorily with comparatively little attention, does not cause any nuisance unless it is too near to occupied buildings or is not well maintained and, unless sewage or effluent has to be pumped, does not need electric or other power.

The small plant is different from that designed for taking flows from towns or from populations of more than about 120 persons, for it has to be able to continue functioning reasonably well even when it is not given the attention that it should receive. All municipal works, even those that have no permanent staff, are regularly visited; sedimentation tanks are sludged, percolating-filter distributors are cleared of stoppages, screenings are removed and sludge beds operated in proper rotation. Even the works serving hospitals and similar institutions can be given sufficient attention because the 'engineer' or handy-man who looks after the hot-water system can make sure that the sewage works perform properly.

As compared with the foregoing, the small sewage-treatment plant serving a private house is liable to be neglected and forgotten and, more often than not, eventually ceases to operate. But it is possible to keep such plant in working condition and obtain effluents of better than Royal Commission standard if the plant is properly designed, inspected not less frequently than once a month and the sedimentation tanks sludged regularly.

The septic tank

Infrequent maintenance is the reason for differences in design between small private sewage works and municipal plant. In particular, in lieu of the municipal sedimentation tank which should be sludged from once to seven times a week, small private works have septic tanks

The septic tank was originated on the fallacious belief that, if sludge were settled and then stored for some months, it would almost completely digest and little or nothing would have to be removed. Accordingly municipal works were provided with tanks intended to work on this principle but, as they invariably became full of sludge which the design rendered difficult to remove, they were eventually either converted to other purposes or demolished.

But a septic tank can be designed merely as a large sedimentation tank capable of storing sludge for a few months and, if it is properly sludged by *complete emptying* at regular intervals, it is satisfactory for removing suspended solids prior to secondary treatment. For this reason the septic tank is almost invariably used when infrequent sludging is inevitable.

When sewage is run through a continuous-flow sedimentation tank of large capacity and sludge and scum are not removed except at intervals of about 3 months, it is found that a thick deposit of sludge collects at the bottom and a thick blanket of scum collects at the top. To ensure that the tank will continue to separate solids

SEWAGE TREATMENT FOR ISOLATED BUILDINGS 241

from the liquid in spite of these deposits, it must have a capacity large enough to permit slow flow of sewage from inlet to outlet between the settled sludge and the blanket of scum for the whole of the time between one sludging and the next. This condition can be ensured by allowing sufficient capacity for sludge settled in a period of 3 months to occupy not more than the bottom third of the tank, scum to occupy not more than the top third and what capacity is left

TABLE 48. *Delivery of chain pumps*

Diameter of barrel (millimetres)	50	63	75	88	100	113	150
Cubic metres raised per hour at 60 revolutions per minute (approximately)	3½	5½	8	11	14	18	32

TABLE 49. *Proportions of septic tanks*

Population served*	Total approximate capacity (cubic metres)	Number of tanks	Dimensions of each tank (metres)		
			length	breadth	depth
20	3·4†	1	2·4	0·8	1·75
40	6·8	2	2·4	0·8	1·75
60	10·2	2	2·86	1·03	1·75
80	13·6	2	3·43	1·14	1·75

* On the basis of sludging at 3-monthly intervals.
† Minimum recommended capacity.

in the middle to be available for free flow of sewage. This condition is well satisfied by providing 0·17 cubic metre of septic-tank capacity for every person served by the works, provided the works have to treat domestic sewage only, with no admixture of trade waste. Table 49 gives recommended proportions of septic tanks to serve various populations of up to eighty persons on the assumption of sludging by *complete emptying* every 3 months. If the sludging period is increased or decreased the number of persons that can be served should be decreased or increased proportionately. It is recommended that, no matter how small the population, the septic tank should never be

Q

smaller than the capacity of the local cesspool-emptying vehicle: it should never be smaller than the 2·73 cubic metres required by Building Regulations.

It is important that the inlet to the septic tank should be so arranged that the sewage enters quietly: there should not be a drop of even as little as 20 millimetres from the invert of the incoming drain as this will militate against settlement. Preferably the invert of the incoming drain should be level with the outlet weir and there should be a dip-pipe of standard catalogue design, or other arrangement, to bring the sewage in below the level of the scum blanket. At the outlet end of the tank should be a weir protected by a scumboard carried down below the bottom of the scum blanket or one-third of the average depth of the tank. The floor of the tank should slope to a sump into which the suction of the tank-vehicle may be dropped. The tank should be covered to prevent nuisance and accident, but it should be ventilated by a free flow of air from inlet to outlet and removable covers should be provided to give access for the suction of the tank-vehicle and for inspection of the inlet and outlet.

A septic tank needs to be completely emptied on sludging, or otherwise the tank-vehicle will take away a load of settled sewage leaving the sludge and scum behind and, as the tank is filling during emptying, any septic tank which is bigger than the capacity of the tank-vehicle must be in two portions so that the flow of sewage can be passed through one while the other is being emptied. The property occupier should make sure that each tank is, in fact, isolated by the closing of a valve or penstock while the emptying is in progress.

If cesspools (or septic tanks) are to be emptied by tank-vehicle, they should be so located that this is possible. Cesspool-emptiers carry in the region of 60 metres of flexible suction pipe: the tank should be near enough to the public highway or other sound road for the suction pipe to reach from the vehicle to the bottom of the tank which must not be lower than 7·5 metres below the top of the cesspool-emptier. As the cesspool-emptier usually stands 2·5 metres above road level, this means that the bottom of the tank must not be more than 5 metres below road level.

Where it is possible to dispose of sludge locally without causing nuisance or contamination of water supplies, the method can be either discharge to trenches which are later filled in and redug, or discharge on to ploughed land which is later reploughed. In both instances the capacity of the trenches or of the furrows must not be less than

SEWAGE TREATMENT FOR ISOLATED BUILDINGS

the entire capacity of the septic tank plus a margin of safety, or there may be spill-over from the land to a stream or some other place where it would be undesirable.

Small percolating filters

The main differences between percolating filters serving very small populations and those at municipal works are in the types of distributors and the forms of beds to suit them. Distributors that have been used at small works have varied from almost completely useless fixed sparge pipes, with no dosing arrangements and buried in the medium, to rotating distributors of very good design and performance. The usual arrangements at present in vogue may be classed under two heads:

1. Rotating distributors.
2. Fixed channels dosed by tipping trays.

Distributors in the former class generally give good distribution. Those in the latter class vary in efficiency according to the design and, on the whole give less even distribution, for which reason they necessitate a larger quantity of filter medium per person served.

Hydraulic head is often limited and, for this reason, small percolating filters may have to be shallow. This again affects the efficiency of the filter and the amount of medium required. Table 50 gives the minimum recommended capacities for small percolating filters according to the type of distributor and the depth of the bed.

TABLE 50. *Minimum capacities of small percolating filters*

Minimum depth (metres)	Minimum cubic metres per person served	
	Revolving distributor	Tipping tray
1·85	0·4	0·6
1·65	0·45	0·675
1·50	0·5	0·75
1·35	0·55	0·825
1·25	0·6	0·9

Until after the Second World War there were no rotating distributors available for those very small percolating filters of the sizes applicable to private houses. This left the choice of a few makes of tipping-tray mechanism. Today Hartley (Stoke-on-Trent) Ltd, make 'Monojet' distributors for circular beds as small as 1·3 metres diameter and there are several designs of rotating distributor that can be used on percolating filters as small as 3 metres diameter. Among the qualities to be looked for when choosing distributors for circular beds is the absence of small-diameter holes in the sparge pipe that can become choked by solids. The Hartley design mentioned above has a separate pipe for each nozzle. For larger beds there is the Ames Crosta Mills 'Simplette' distributor which consists of an open channel with V-notch outlets. This is rotated by a water-wheel mechanism. It was with the 'Simplette' distributor and record of humus by grassland that the author obtained the highest quality final-effluent analyses that he has ever seen (suspened solids nil and B.O.D.$_5$, 3 milligrammes per litre).

Distributors for the smaller beds are often 'over-fed', that is the feeds pass over the surface of the bed, as in the 'Monojet' and the 'Simplette'. Distributors for beds larger than 5·5 metres diameter are often 'under-fed', as are those for municipal works, the feed pipes being carried through the medium. All except those which are caused to rotate by water-wheel or some similar mechanism, but which are driven by the reaction of the jets, need to have a dosing siphon. In some makes this is incorporated in the central column of the distributor: if there is no such arrangement a dosing tank and low-draught siphon must be provided (see Chapter 15).

Tipping-tray distributor mechanisms are manufactured by Tuke & Bell Ltd, for rectangular beds measuring from 2·14 by 0·76 to 7·7 by 3·05 metres or by William E. Farrer Ltd, for beds measuring 1·52 by 0·92 to 6·1 by 2·74 metres. These distributors consist of a tray which, when filled with settled sewage, tips over, causing a flush that fills a series of cast-iron or steel channels arranged to distribute the flow over the bed. The smaller beds have a one-way tipper at one end but the larger beds have a central tipper which alternately distributes in opposite directions to the two halves of the bed.

Percolating filters for small works should be constructed generally on the lines of those for larger works. They should have ample ventilation at the bottom. The medium should be coarse, having particles of about 60 millimetres diameter as a precaution against choking of the bed should sludging of the septic tank be neglected. The side walls should be carried above the top of the medium to protect the filter from the weather except where the height of the wall is determined by the type of distributor. On no account should percolating filters be covered by slabs or boards because this interferes seriously with ventilation and encourages neglect, but where the filters are under trees, wire-mesh covers should be provided as a protection against falling leaves.

Humus removal

It has been found that in most, but not all, instances where humus tanks have been installed at small works, they have been neglected to such an extent that they have become filled with humus and completely ineffective. Humus tanks *can* be used in those rare circumstances where pumping of crude sewage is necessary, for then the humus can be drawn off continuously to the suction well. In most cases it is much more satisfactory to remove humus not by settlement in tanks but by irrigating the percolating-filter effluent over grassland, about 3 square metres being allowed per person served, and collecting the final effluent in a ditch leading to a catch-pit with an outlet to the nearest watercourse.

Activated-sludge works and packaged plants

Various types of 'packaged plants' are available. These include very small normal activated-sludge works and also activated-sludge works with 'extended aeration', that is, aeration is continued at a high rate and long enough to effect a considerable degree of aerobic digestion of organic solids. These latter methods require a much greater input of electric power than the normal activated-sludge plant.

A new method of treatment is the Ames Crosta Mills BioDisc process. This consists of a steel tank very similar to an Imhoff tank* but containing a number of discs of expanded metal which are mounted on a common longitudinal axle and slowly rotated so that

---

* An Imhoff tank is a type of sedimentation tank used in Germany and America but seldom in Great Britain. It has an upper compartment that serves as sedimentation tank connecting via a slot in the floor to a lower sludge-storage and digestion tank.

they alternately pass through and aerate the settled sewage, then rise to dissolve air. The discs soon become coated with organic film. The method produces a Royal Commission effluent.

Land treatment

Treatment of settled septic-tank effluent on land near residences is generally not to be recommended because it can be offensive and because of the flies the method encourages. Furthermore, irrigation areas, unless properly maintained by resting, rotating and adjusting so that the sewage is adequately distributed, will not give satisfactory effluents.

A method that has been much used but should not be encouraged is sub-irrigation. This is the distribution of settled sewage to a system of underdrains by means of a dosing tank and siphon. It has been claimed that this treats the sewage but, in fact, it can be little more than illicit discharge of untreated sewage to the subsoil. Although the method has its uses overseas or in very remote places, in most circumstances it should not be permitted.

Disposal of final effluent

After humus removal the final treated effluent may be discharged to a watercourse, but where there are no watercourses because the subsoil is porous, it is unavoidable that the effluent must be soaked into the ground. This can be troublesome for, although all rainfall soaks away in limestone and some other similar districts and soakaways can be satisfactorily installed to dispose of surface water, sewage-works effluents contain suspended solids and encourage organic growths which will choke soakaways in a comparatively short time. In these circumstances it is permissible to use sub-irrigation, not as a means of treatment but for the disposal of a thoroguhly treated effluent. In chalk and other limestone districts it will often be found that a line of unjointed underground pipework will find a natural joint in the rock where the effluent will flow away freely. Also, when the treated effluent is irrigated over the surface of the land it may soak away. It is suggested that, where possible in limestone areas, the percolating-filter effluent should be irrigated over the ground, collected in a ditch and finally discharged to a deep soakaway.

Pumping

Pumping is often required at small sewage works because of insufficient fall to permit the proper depth for a percolating filter or to lift from the drains to sewage-treatment plant uphill of premises that are on the high side of the road.

Wherever possible the pumping should be from the septic tank to the percolating filter, not from the drains to the septic tank. If it is unavoidable that pumping must be from the drains to the septic tank it becomes necessary to have pumps capable of passing crude-sewage solids. Then it is best to have the smallest size sewage-pumping station as described in Chapter 3 or an ejector station as described in Chapter 4, and either of these could cost more than the treatment plant. Furthermore, the delivery of pumps large enough to pass sewage solids and keep a 100-millimetre rising main clean will unduly disturb the contents of the septic tank unless it is a large one.

If it is possible, as is often the case, for the drains to gravitate to a suitably located septic tank and for the tank-effluent to be pumped to the percolating filter, the pumping plant adds little to the cost of the works, for pumps for settled sewage need not be larger than necessary to take the flow and to keep a rising main of, maybe, 38 millimetres internal diameter clean: this is a discharge of about 0·055 cubic metre per minute.

When this is done the pump, rising main and filter distributor should be capable of discharging 0·055 cubic metre per minute or the maximum rate of flow from the premises (say three times the average daily water demand) whichever be the greater. The automatic float gear should be adjusted and the capacity of the suction well so designed that the quantity delivered at any one time is not excessive: a suitable figure is 10 litres per square metre of filter-bed surface. The intermittent pumping will do away with the necessity for any dosing mechanism in those circumstances where otherwise an external dosing tank would be necessary. A little surcharge of the suction well above cut-in level of the pump will take care of short spells of discharge exceeding three times dry-weather flow.

Vertical-spindle submersible automatic electric pumps with attached float gear are available in several makes and sizes. A small house of brick or timber construction should be built above them. The rising main should be given adequate protection against frost,

Appendix

Imperial and metric equivalent units

Linear

25·4 millimetres	= 1 inch
3·28084 feet	= 1 metre
1·609344 kilometres	= 1 mile

Square

6·4516 square centimetres	= 1 square inch
10·7639 square feet	= 1 square metre
1·19599 square yards	= 1 square metre
2·47105 acres	= 1 hectare
2·58999 square kilometres	= 1 square mile

Cubic

3·78531 litres	= 1 U.S. gallon
4·54596 litres	= 1 Imperial gallon
35·3147 cubic feet	= 1 cubic metre
264·179 U.S. gallons	= 1 cubic metre
219·975 Imperial gallons	= 1 cubic metre

Weight

2·20462 pounds	= 1 kilogramme

Pressure

14·6959 pounds per square inch	= 1 atmosphere
1·03323 kilogrammes per square centimetre	= 1 atmosphere
10 metres head of water	= 1 kilogramme per square centimetre
14·2233 pounds per square inch	= 1 kilogramme per square centimetre

6894·76 newtons per square metre = 1 pound per square inch

Force
1 newton = 1 kilogramme metre per second per second

Temperature: *Fahrenheit, Centigrade* (*Celsius*) *and Kelvin*

(°F − 32)/1·8 = °C
°K − 273·15 = °C

Heat and Energy

1055·06 joules (1·05506 kilojoules) = 1 British thermal unit
4186·8 joules = 1 kilocalorie
1·35582 joules = 1 foot pound
1 joule = 1 newton metre
3088·03 foot pounds = 1 kilocalorie
426·935 kilogramme metres = 1 kilocalorie
3·96832 British thermal units = 1 kilocalorie
859·845 kilocalories = 1 kilowatt hour

Power

101·972 kilogramme metres per second = 1 kilowatt
6·11832 cubic metres of water metres per minute = 1 kilowatt
1000 (1 kilojoule) joules per second = 1 kilowatt
1·34102 horse-power = 1 kilowatt
75 kilogramme metres per second = 1 metric horse-power
1·01387 metric horse-power = 1 horse-power

Bibliography

1. ANDERSON, NORVAL E. 'Design of Final Settling Tanks for Activated Sludge', *Sewage Works Journal*, January 1946.
2. BABBITT, H. E. *Sewerage and Sewage Treatment*, 1953.
3. BILHAM, E. G. 'Classification of Heavy Falls in Short Periods', *British Rainfall*, 1935.
4. COPAS, B. A. 'Storm Water Storage Calculations', *Journal Institution of Public Health Engineers*, Vol. LVI, Part 3, July 1957.
5. ESCRITT, L. B. and RICH, S. F. *The Work of the Public Health Engineer*, 1959.
6. ESCRITT, L. B. and YOUNG, A. J. M. 'Economic Surface-water Sewerage: A Suggested Standard of Practice', *Journal Institution of Public Health Engineers*, Vol. LXII, Part 4, October 1963.
7. FEDERATION OF SEWAGE WORKS ASSOCIATIONS. 'Occupational Hazards in the Operation of Sewage Works', *Manual of Practice*, No. 1, U.S.A. 1944.
8. FEDERATION OF SEWAGE WORKS ASSOCIATIONS. 'Utilisation of Sewage Sludge as Fertiliser', *Manual of Practice*, No. 2, U.S.A. 1946.
9. GLASSPOOLE, J. 'The Areas Covered by Intense and Widespread Falls of Rain', *Proceedings Institution of Civil Engineers*, Vol. 229, Part 1, 1929/30.
10. GREELEY, S. A. and STANLEY, W. E. Section 22, Sewerage. *Handbook of Applied Hydraulics*, 1952.
11. HART, C. A. 'Correspondence on Rainfall and Run-off', *Journal Institution of Municipal and County Engineers*, Vol. LIX, No. 18, p. 978.
12. IMHOFF and FAIR. *Sewage Treatment*, 1956.
13. LLOYD-DAVIES, D. E. 'The Elimination of Storm-water from Sewerage Systems', *Proceedings Institution of Civil Engineers*, Vol. CLXIV, 1906.
14. MILLS, E. V. 'The Treatment of Settled Sewage in Percolating Filters in Series with Periodic Changes in Order of the Filters: Results of Operation of the Experimental Plant at Minworth, Birmingham, 1940 to 1944', *Journal Institute of Sewage Purification*, 1945, Part 2.
15. MILLS, E. V. 'The Treatment of Settled Sewage by Continuous Filtration with Recirculation of Part of the Effluent', *Journal Institute of Sewage Purification*, 1945, Part 2.
16. MINISTRY OF HEALTH. '*Accidents in Sewers: Report on the Precautions Necessary for the Safety of Persons Entering Sewers and Sewage Tanks*', 1934.

17. MINISTRY OF HEALTH DEPARTMENTAL COMMITTEE ON RAINFALL AND RUN-OFF. *Journal Institution of Municipal and County Engineers*, Vol. LVI, No. 22, p. 1172.
18. MINISTRY OF HOUSING AND LOCAL GOVERNMENT. *Methods of Chemical Analysis as Applied to Sewage and Sewage Effluents*, 1956.
19. ORMSBY, M. T. M. 'Rainfall and Run-off Calculations', *Journal Institution of Municipal and County Engineers*, Vol. LIX, No. 16, p. 889.
20. RILEY, D. WEARING. 'Notes on Calculating the Flow in Surface Water Sewers', *Journal Institution of Municipal and County Engineers*, Vol. LVIII, No. 20, p. 1483.
21. ROYAL COMMISSION ON SEWAGE DISPOSAL, *Fifth Report*, 1908.
22. ROYAL COMMISSION ON SEWAGE DISPOSAL, *Eighth Report*, 1918.
23. SCOBEY. 'Flow of Water in Concrete Pipes', *U.S. Dept. Agr. Bull.* 852, 1902.
24. SWANWICK, J. D., SHURBEN, D. G. and JACKSON, S. *A Report on the Water Pollution Research Laboratory Survey of the Performance of Sewage Sludge Digesters throughout England, Scotland and Wales.*

Bibliographical references in the text are shown thus: [10].

Index

Acidity, 122
Activated Sludge Ltd, 217
Activated sludge, 185, 215
   detention period, 218
   recirculation, 221
   systems, 135
   treatment, 214
   works, formula, 216
   works, small, 245
Aerated detritus channels, 150
Aerating cone, 223
Aeration tanks, surface area, 219
Aerodrome drainage, 46
Air blowers, efficiencies, (table) 73
Air: compression of, (table) 70
   filters, 221
   lifts, 71
   lifts, performance, (table) 72
   mains, sizes, (table) 220
   supply, 219
   test, 85, (table) 87
Algae ponds, 230
Alkalinity, 122
Alternating double filtration, 209
Ames Crosta Mills & Co. Ltd, 185, 196, 223, 244, 245
Ammoniacal nitrogen, 122
Asbestos-cement pressure pipes, 79
Automatic pumping stations, 58
Automatic sampler, (figure) 126
Axial-flow pumps, 48

Backdrop, 92
Balancing flows, 161
Beckton Sewage-treatment Works, 135, 192, 199
Bilham's formula, 31
Bio-aeration, 224
Biochemical oxygen demand, 123
BioDisc plant, 203, 245
Bio-filtration, 224
Bio-flocculation, 224
'Bleeding off' sludge, 170
B.O.D.s, kg per head per day, (table) 128
Brick sewers, maximum depth of, (table) 82
Bricks, 93
Broad irrigation, 137
Bruges, C. Ernest, 18
Building Regulations, 234, 237, 238

Bulking, 225

Cascades, 93
Cast-iron pipes, safe span of, 80
Cast-iron segmental sewers, 83, 84
Cast-iron tubbing, 95
Cavitation, 51
Centrifugal pumps, 48
Centrifugal pump characteristics, 49
Cesspool, 237
   emptying, 242
   emptying vehicles, 238
Chain pumps, 239
   delivery of, (table) 241
Chambers, construction of, 93
Change of impermeability, 36
Chemical closets, 234
Clay pipe flexible joints, (table) 76
Clearance between sewers, (tables) 22, 23
Coastal problems, 100
'Coilfilter', 182
Cold digestion, 188
Colloids, 164
Combined sewers, capacities of, 27
Combined system, 24
Comminutors, 143
Compression of air, (table) 70
Compressors, 221
Conservancy sanitation, 233
Constant-velocity detritus channels, 150
Contact beds, 202
Continuous-flow sedimentation, 165
Crimp, W. Santo, 18
Crimp's formula for sludge dewatering, 177
Crossness Sewage-treatment Works, 127, 134, 135, 152, 177, 214, 223

Dangerous gases in sewers, (table) 97
Danson, P. S. S., 215
Decanting valves, 198
Detention period, economic, 167
Detritus channels, aerated, 150
   constant-velocity, 150
Detritus: quantity of, 153
   settlement, 149
   tanks, 143
Diameters of sewers, minimum, 21
Differential float gear, 146
Diffused-air method, 214, 217
Diffusers, 217

INDEX

Digestion process, heat produced by, 191
Digestion tanks, heat losses from, 192
Discharge of orifices, 159
Discharge over weir, 161
Disintegration, 147
Dissolved Air Flotation Thickening Process, 185, 222
Dissolved solids, 123
Distributors for percolating filters, 212, 213
Dome diffusers, 218
Dorr Detritor, 149
Dorr-Oliver Co. Ltd, 182, 222
Dorr-Oliver 'B' type heater, 195, 196
Dosing tank, 212
Drainage area diagram, 42
Drop manholes, 92
Dry-weather flow defined, 26
Dry well, 51

Earth pressure, 81
Economic detention period, 167
Economic diameters of pipework and valves, (table) 55
Economic velocity in rising main, 68
Eddy bucket, 166
Efficiencies of electric motors, (table) 65
Efficiencies of air blowers, (table) 73
Effluent polishing, 228
Effluents: standards of, 128
 treatment on land, 230
Ejectors, sewage, 69
Electrodes, 58
Electrostatic filters for air, 221
Enclosed filtration, 209
Escritts' Tables, 18, 175
Excess-gas burner, 200

Faecal matter, 121
Fair, Gordon M., 148
Farrer Ltd, William E., 244
Filter media, properties of, 204
Final effluent disposal, 246
Fittings, loss of head through, (table) 63
Flat-rate calculations of rainfall run-off, 44
Float gear, differential, 146
Float switches, 58
Floating roof, 196
Floats, 101
Flocculation, 164
Flushing of sewers, 96
Foaming, 226
Foxboro-Yoxall Ltd, 132

Gas collection, 199
Gas discharge through orifices, (table) 200
Gas production, 189
Gasholder capacity, 199
Gasholder roof, 198
Glenfield & Kennedy Ltd, 229
Gradients, 19
 for sludge mains, (table) 140
 minimum, (table) 20
Granular fill, 79
Gratings, 92
Grit washer, 149

Handwheels, 56, 57
Hartley (Stoke-on-Trent) Ltd, 244
Hazen's theory of settlement, 166
Heat exchange calculations, 192
Heat losses from digestion tanks, 192
Heat produced by digestion, 191
Heat treatment of sludge, 184
'Heatamix' system, 196
Heater for sludge, 'B type', 195
 'Simplex', 195
High-intensity cone, 223
House connexions, 98
Housing, impermeability factors, 31
Humus, 176, 185
 removal, 245
Hydraulic tests, 86
Hydrograph method, 46

Imhoff, Karl, 148
Imhoff tank, 245
Impermeability: change of, 36
 factor, 29, (table) 30, 31
Impermeable area, 29
Incremental feeding, 224
INKA system, 222
Inspection chambers, 89
Intercepting traps, 98
International Organization for Standardization, Recommendation R.13, 80
Inverted siphons, 114
 testing, 86
Isolated buildings, sewage treatment for, 232

Jenkins, S. S., 215
Jenning's joinders, 99
Joints: ogee, 78
 socketed, 78, 114

Kessener system, 224
Komline-Sanderson 'Coilfilter', 182
kVA charges, 64

Ladders, 91
Lamphole covers, 92
Land treatment, 134, (table) 136, 137, 246
 of effluents, 230
Lloyd-Davies calculation, 32, (table) 34
Lloyd-Davies formula, 27
Lloyd-Davies method, 27
Loans, repayment of, (table) 66
Loss of head through fittings, (tables) 63, 64
Lunar clock, 103

Magnetic flow meter, 132
Malleable cast-iron pipes, 79
Manholes, 21, 89
 construction of, 93
 covers, 92
 covers, ventilating, 98
 drop, 92
 inverts of, 90
 ironwork, 90
 pre-cast concrete, 95
 wall thickness, (table) 94
Mass concrete, 95
Maximum velocities, 19

254 INDEX

McGowan's formula, 123
Mechanical agitation, 222
Mechanized sedimentation tanks, 172, 174
Mesophilic digestion, 189
Mesophilic tanks, 197
Metric (nominal) dimensions, 75
Micro-strainers, 229
Minimum diameter for sewer, 21
Minimum gradients, (table) 20
Ministry of Health formulae, 32
Mixed-flow pumps, 48
Mixed liquor, 215
Mogden, 197
   formula, 138
'Monojet' distributors, 244

Nitrates and Nitrites, 122

Ogee joints, 78
Optimum diameter of rising main, example calculation, (table) 65
Organic content of sewage, 121
Organic film, area required to treat sewage, 203
Orifices, discharge of, 145, 159
Oxygen absorbed, 123

Packaged plants, 245
Partially separate system, 24
Particles, settlement of, (table) 143, 163
Percolating filter schemes, 135
Percolating filter treatment, 202
Percolating filters, arrangement of, 207
   capacities of, 205
   construction of, 210
   depth of, 208
   design loads for, 207
   distributors for, 212, 213
   flow distribution, 211
   loads on, (table) 206
   small, (table) 243
   surface loading of, 208
Persons per house and water service, 27
pH value, 122
Picket-fence thickener, 164
Pipe sewers, 78
Pipelines, pressure, 79
Pipes: asbestos-cement, 78
   asbestos-cement pressure, 79
   cast-iron, 79
   malleable cast-iron, 79
   pre-stressed concrete, 78
   salt-glazed ware, 78
   steel, 79
   supported on piers and under heavy earth load, 81
   vitreous-enamelled fireclay, 78
   vitrified clay, 78
Pipework and channels, 139
Pipework and valves for pumping stations, 54, (table) 55
Pneumatic pump-starting gear, 58
Porteous process, 184
Power calculations, pumping, 63
Power factors of electric motors, (table) 65
Pre-cast concrete manholes, 95

Pre-cast concrete segmental sewers, 83
Preliminary treatment, 142
Pressure, external water, 81
Primary digestion, heat requirements, 191
Proportional depth, area, velocity and discharge of pipes flowing partly full, (table) 61
Psychoda, 204
Public Health Act, 1936, 233, 237
Pumping at small sewage works, 247
Pumping directly from sewer, 60
Pumping stations: automatic, 58
   capacities, 62
   layout, 52
   pipework and valves, 54, (table) 55
Pumps: axial-flow, 48, 49
   cellar-emptying, 57
   centrifugal, 48
   centrifugal, characteristics, 49
   dimensions of, 53
   mixed-flow, 48, 49
   reciprocating, 49
   submersible, 51
   turbine, 49
   vertical-spindle, 51
   volute centrifugal, 48
Pyramid, capacity of, 171
Pyramidal-bottomed tanks, 171

Quiescent sedimentation, 165

Rainfall intensities, 31, (table) 33
Ramps, 92
'Rational' method, 27, 33
   calculation, 37, (table) 38
   simplified, (table) 40
Recirculation, 209
Reinforced concrete, 95
Repayment of loans, (table) 66
Rheostat starters, 59
Riley, D. Wearing, 42
Rising mains: economic velocity, 68
   example calculation of optimum diameter, (table) 65
   testing, 86
Rotary blowers, 221
Royal Commission on Sewage Disposal, 123, 127, 128, 137, 143, 154, 205
Run-off coefficients, (table) 37
Run-off per roofed or paved hectare, (table) 39

Saddle, 99
Safety harness, 92
Salt-glazed ware pipes, 78
Sampler, automatic, (fig.) 126
Samples, weighted, 125
Sampling, 124
Scobey, F. C., 18
Screening, manual, 142
Screenings, 147, 148
Screens, mechanically raked, 144
Scum-removal device, 198
Sea, discharge of sewage to, 101
Sea outfalls, 101, 105, 114
   discharge coefficients, (table) 107
   stage-by-stage calculation, 112
Sea wall, 113

INDEX 255

Sedimentation, 163
  formula for, 165
  tank capacities, 170
  tanks, details, 175
  tanks, inlets, 167
  tanks, mechanized, 172, 174
  tanks, pyramidal, 171
  tanks, types of, 168
Self-cleansing velocities, 19
Separate system, 24
Separation of storm water, 154
Septic tanks, 202, 240, (table) 241
Settlement speeds of particles, (table) 143
Sewage: defined, 121
  discharge to sea, 100
  ejector, 69
  flows, 26
  nature of, 121
  organic content, 121
  sampling, 124
  soil, 25
  strength of, 121, 123, 127
Sewage treatment for isolated buildings, 232
Sewage-treatment works, 135
  cost, 135
  land for, 134
  layout, 139
  siting of, 133
Sewage works, small domestic, 239
  small, cost of, 232
  small, pumping at, 247
Sewerage systems, 24
Sewers: above ground level, 116
  cast-iron segmental, 83
  construction of, 75
  dangerous gases in, (table) 97
  definition of, 21
  flow in, 17
  flushing of, 96
  general requirements for, 21
  in tunnel, 83
  large-diameter, 80
  location of, 22
  maximum depth of, (table) 82
  pipe, 78
Sheffield system, 224
'Simplette' distributor, 244
'Simplex' sludge heater, 195
'Simplex' system, 223
Siphon: flushing, 96
  inverted, 114
Sludge: age, 226
  as manure, 185
  'bleeding off', 170
  content of, 177
  dewatering, formula for, 177
  digestion, 187
  disposal, 176, 178
  disposal costs, (table) 179
  drying beds, 179
  flow, 140
  freezing, 184
  gas, content and calorific value, 190
  heat treatment of, 184
  hoppers, 169
  incineration, 184
  liquor pumping station, 180
  mains, gradients for, (table) 140
  methods of heating, 195
  quantities of, 176
  shipping to sea, 183
  specific gravity of, 178
  specific heat of, 178
  utilization, 184
  -volume index, 225
Sludging, 169
Soakaway systems, 47
Soakaways for effluent, 246
  construction of, 93, 96
Socketed joints, 78
Soil sewage, 25
  rate of flow, 26
Soil sewer calculation sheet, 28
Soil sewers, capacities of, 25
Solids reduction, 189
Specific speed, 50
Spindles, pull on, 58
Square roots, (table) 108
Stage-by-stage outfall calculation, 112
Stand pipe, 57
Standing-wave flume, 130
Starting gear, 59
Starting times of electric pumps, (table) 61
Station finder, 102
Steel pipes, 79
Step-irons, 91
Storage tanks: coastal, 105
  construction of, 113
Storm overflows, 93
Storm sewage defined, 121
Storm tanks: design of, 158
  function of, 156
  spill-over from, (table) 160
  theoretical capacities, 157
Storm water: separation and treatment of, 154, 159
  storage of, 62
Strength (McGowan), 123
Strength of sewage, 121, 127
Sub-irrigation, 246
Submersible pumps, 51
Suction well, dimensions of, 59
Sullage: defined, 121
  disposal, 236
Surface aeration, 214, 222
Surface coefficients, (table) 193
Surface drainage, special cases, 46
Surface water defined, 121
Surface-water sewer capacities, 27
Surplus activated sludge, 215, 221
Suspended solids, 123
Systems of sewerage, 24

Tangent method, 41
Tank sewers, 113
Tapered aeration, 224, 225
Testing sewers and rising mains, 85
Tests, air and water, (table) 87
Thermal conductivities, (table) 194
Three-point problem, 102
Three-throw pumps, 49
Tidal experiments, 101
Tide curve, 112
Tide flap, 101
Tide recorder, 112

Time–area graph, 42, (figure) 45
Time to empty tank, 106
Tipping-tray distributors, 244
Trade wastes: charges for, 138
    treatment of, 138
Tuke & Bell Ltd, 244
Tumbling bays, 93
Tunnel, sewers in, 83
Turbine pumps, 49
Turned and bored joints, 114

Vacuum filtration, 181
Valves, time to open, 57
Velocities in sewage-works conduits, 140
Velocities, maximum, 19
Velocities, self-cleansing, 19
Ventilating columns, 98
Ventilating manhole covers, 98
Ventilation, 97

Vertical drops, 92
Vibrating, 95
Vitreous-enamelled fireclay pipes, 78
Vitrified clay pipes, 78

Walls, thickness of manhole, (table) 94
Washout pumping station, 181
Water cushions, 93
Water demands, 26
Water Pollution Research Laboratory, 178
Water pressure, external, 81
Water test, 85, (table) 87
Wave height, 113
Weighted samples, 125
Weir, discharges over, 161
Weirs, thin-plate, 130

Zimmerman process, 184